Lessons in Bioinformatics - Dot Plots

Lessons in Bioinformatics, Volume 1

Björn Olsson

Published by Björn Olsson, 2025.

LESSONS IN BIOINFORMATICS - DOT PLOTS

First edition. May 22, 2025.

ISBN: 978-9199033013

Written by Björn Olsson.

Table of Contents

What is a dot plot and how is it calculated? 3

Separating relevant matches from random noise using the sliding window approach 7

Scoring nucleotide matches with substitution matrices 12

Using dot plots to compare homologous genes 17

Using dot plots to compare homologous protein sequences 21

Plotting a protein sequence against itself to find repeats and regions with compositional bias 26

Domain structure in protein sequence dot plots 35

Advanced algorithms for fast dot plot comparison of genome sequences 37

Ultra-fast algorithms for analysis of large genomes 47

Application example 1: Comparing multiple clustered genomes 54

Application example 2: Identifying gene duplication events 59

Application example 3: Rearrangement hot spots in cow genomes 63

Application example 4: Structural differences between rice genomes 66

Application example 5: Measuring the overall genome sequence similarity 73

A brief history of research on dot plot methods 81

Bibliography 90

Preface

This book is intended to be the first volume in a series of books, *Lessons in Bioinformatics*, on fundamental concepts and methods in bioinformatics. Each book will focus on a single topic, which is explained in a comprehensive and detailed, yet easy-to-read, format. The books are primarily addressed towards beginner bioinformaticians who are building up a broad base of knowledge and skills. Previous knowledge of bioinformatics is not assumed, so everything covered in this text should be easy to understand even if it is the first book you read on bioinformatics.

The current volume covers the topic of dot plots, starting with a basic introduction to what dot plots are and how a simple dot matrix is generated. The text thereafter gradually progresses towards more advanced methods with numerous examples explaining each new concept that is introduced, for example, the use of sliding windows and substitution matrices. It shows how dot plots can be used to identify inserts and deletes, repeat regions, compositional bias, and domain structure in protein sequences.

When applied to the study of genome sequences, the algorithms used by modern dot plot software are more complex, since analysis of very long sequences – consisting of up to hundreds of millions or even a few billion symbols – is far too slow unless the algorithms are cleverly designed to be very fast. Recently published dot plot software for genome sequence analysis therefore relies on advanced concepts such as position trees, suffix arrays and minimizers. The book attempts to explain these algorithmic ideas in a reasonably easy way, to make them accessible to readers regardless of academic background.

The final chapters of the book present application examples from research articles and show step-by-step how the dot plots shown in the journal articles can be reproduced by the reader. At the very end, the final chapter gives an overview of the history of research on dot plot algorithms, from the 1970s to the present day. This gives a historical perspective of how algorithms have advanced from the time when short sequences of a few hundred symbols were painstakingly analyzed by the slow computers of that time, or even by hand with pen and paper, to the powerful dot plot tools used today, that can compare medium-sized genomes in a matter of seconds or two eukaryote genomes containing over a billion symbols in a few minutes.

If you find this book useful, then look out for upcoming titles in the same series, *Lessons in Bioinformatics*, which are intended to cover the fundamental topics that every bioinformatician should master.

If you find any errors, omissions, or other shortcomings in this book, then please consider sending your feedback to the author at bjorn.bioinformatics@gmail.com so that corrections and improvements can be made for future editions.

What is a dot plot and how is it calculated?

One of the basic, and most often performed tasks in bioinformatics, is to compare two biological sequences to each other. The sequences can be of different kinds, and there can be many different purposes for making the comparison. The two sequences we compare to each other may be the coding sequences from two genes, the mRNA sequences from two transcripts, the amino acid sequences of two proteins, the nucleotide sequences of two chromosomes, or even two full-length multi-chromosome genome sequences from two eukaryote organisms. A conceptually and computationally simple method for such pairwise comparisons is a so-called dot plot. The dot plot approach is easy to understand, comparatively easy to implement, and fast to execute. Yet, despite its simplicity, the dot plot approach is quite powerful in terms of the amount of insight it can produce, and it is very versatile in terms of its range of applications. Also, as we will see in later chapters, when sequences are very long, the algorithms needed to produce a dot plot in reasonable time can be quite complex.

A dot plot is a two-dimensional visualization of the relatedness, or shared content, of two biological sequences. To explain how dot plots are generated, we will first go through a very simple example. After that, we will move on to gradually more complex variations of the basic method. Briefly summarized, a dot plot is created by first making a two-dimensional matrix where the columns represent one sequence, and the rows represent another sequence. The matrix cells representing positions where the two sequences contain identical symbols are then filled with “dots”. This can be easily done by going through the matrix row by row and placing a “dot” in every cell where the symbol (nucleotide or amino acid) in the corresponding column is the same as the one that the row represents. **Figure 1** shows a simple example of this process.

		1	2	3	4	5	6	7	8	9	10	11	12	13	14	15
		A	T	G	A	C	A	C	A	G	C	T	T	C	A	G
1	A	X			X		X		X						X	
2	T		X									X	X			
3	G			X						X						X
4	A	X			X		X		X						X	
5	C					X		X			X			X		
6	A	X			X		X		X						X	
7	C					X		X			X			X		
8	A	X			X		X		X						X	
9	A	X			X		X		X						X	
10	C					X		X			X			X		
11	T		X									X	X			
12	T		X									X	X			
13	C					X		X			X			X		
14	A	X			X		X		X						X	
15	G			X						X						X

Figure 1. Dot matrix comparison of two nucleotide sequences. One sequence is laid out horizontally and the other vertically. In the matrix, we mark every cell that represents identical nucleotides in the two sequences with a "dot" (in this illustration the letter X).

For the example dot matrix in **Figure 1**, we have placed one sequence horizontally, so that its nucleotides are represented by columns 1, 2, etc., until column 15. The other sequence was placed vertically, so that its nucleotides are represented by rows 1 to 15. When filling in the matrix we can of course use any symbol to represent the "dots" for the dot plot, and in this example, we have used the letter X. The algorithm consists of an iteration (repetition) going over the rows 1 to 15, and for each row going through all columns from 1 to 15. Starting with row 1, we will then see that the nucleotide A, i.e. adenine, written at the beginning of that row, also occurs in columns 1, 4, 6, 8 and 14, representing all adenines in the horizontal sequence. Then going through row 2, we check for rows representing thymine (T) in the horizontal sequence, and thereby fill in the cells in columns 2, 11 and 12 with X symbols.

For a basic dot plot tool, this simple procedure is all that is needed to generate the data for the plot, so that the tool can proceed to create a figure that will be the actual plot shown to the user. Note that to avoid confusion, we should distinguish between the dot **matrix**, as the one shown in **Figure 1**, and a dot **plot**, which is the graphical visualization shown to the user as output from the tool. The data from a given dot matrix can be visualized in different ways in the dot plot, as we will see later when discussing examples of dot plot programs.

To formalize the procedure described in the previous paragraphs, we define a function $F(i,j)$ to calculate the score for any cell (i,j) in the matrix and a function $s(x_i,y_j)$ to compare the nucleotides at positions i and j in the two sequences x and y. For any cell in the matrix, we can then set a score of either 1 or 0, depending on whether the two sequences have identical or non-identical nucleotides in the corresponding position, as follows:

$$F(i,j) = s(x_i, y_j) \qquad Eq\ 1$$

where

$$s(x_i, y_j) = \begin{cases} 1\ (x_i = y_j) \\ 0\ (x_i \neq y_j) \end{cases} \qquad Eq\ 2$$

In accordance with *Eq 1* and *Eq 2,* we have written an X in every cell in the matrix in **Figure 1** where $s = 1$ and left the cell blank where $s = 0$. For example, the cell $(1,1)$ is marked with X since both sequences begin with A, i.e. $x_1 = \mathrm{A}$ and $y_1 = \mathrm{A}$. Similarly, cell (2,1) was left blank since $x_2 = \mathrm{T}$ and $y_1 = \mathrm{A}$. Since each cell is independent of all other cells, it does not matter in which order we set the scores for cells. We can apply *Eq 1* to all cells row by row, column by column, or even in random order, provided that all cells are included in the procedure.

When looking at the resulting matrix in **Figure 1**, we can make some observations about the sequences that are being compared. The first observation is illustrated in **Figure 2**. As the highlighting of matrix cells in this figure shows, there is an almost complete diagonal of "dots" between the northwest (upper left) corner and the southeast corner. Along this diagonal, only one X is missing for the diagonal to be complete. This shows that the two sequences are almost identical. Obviously, if the sequences were completely identical, we would see a contiguous diagonal all the way from one corner to the other.

		1	2	3	4	5	6	7	8	9	10	11	12	13	14	15
		A	T	G	A	C	A	C	A	G	C	T	T	C	A	G
1	A	X			X		X		X						X	
2	T		X									X	X			
3	G			X						X						X
4	A	X			X		X		X						X	
5	C					X		X			X			X		
6	A	X			X		X		X						X	
7	C					X		X			X			X		
8	A	X			X		X		X						X	
9	A	X			X		X		X						X	
10	C					X		X			X			X		
11	T		X									X	X			
12	T		X									X	X			
13	C					X		X			X			X		
14	A	X			X		X		X						X	
15	G			X						X						X

Figure 2. Dot matrix comparison with the main diagonal highlighted with grey background colour, showing that the two sequences are almost identical, with the exception being an A/G substitution at (9,9).

In the case shown in **Figure 2**, we may conclude that the two sequences are highly likely to have a common biological origin, and that one of the sequences has undergone a point mutation, where either G was substituted with A in one sequence, or A was substituted with

G in the other sequence. What this example illustrates is also that a dot plot can be used as a quick, but not very precise, way of estimating approximately how closely (or distantly) related two sequences are. The rule of thumb would be that for pairs of closely related sequences, the dot plot will show a clearly discernible diagonal along a significant part of the lengths of the two sequences. For a more precise estimate of the evolutionary distance between the sequences, we would need a numerical measure of similarity, as will be discussed later.

Separating relevant matches from random noise using the sliding window approach

In **Figure 2** we had manually selected the diagonal that is most informative about the relatedness of the two sequences and highlighted it to facilitate the interpretation. For a dot plot tool to do this automatically, there needs to be a general method for identifying diagonals of relevance. **Figure 3** shows an example of such a method, where all contiguous diagonals of length 3 or longer are highlighted. We can here see that in addition to the two main diagonals, stretching from matrix cells with coordinates (1,1) to (8,8) and, after the A/G substitution at (9,9), from (10,10) to (15,15), there are also shorter diagonals, such as the one stretching from cell (4,6) to (6,8) and from (7,13) to (9,15). These diagonals represent shorter stretches of contiguous identical nucleotides. In the first case, i.e. (4,6) to (6,8), both sequences contain the subsequence ACA, and in the second case, (13,7) to (15,9), both contain CAG. There are also some short diagonals in the opposite, northeast-southwest orientation, such as the one from (3,15) to (5,13), which represent reverse matches, in this case GAC in the horizontal sequence matching in reverse with GAC in the vertical sequence. There are also cases of palindromic matching sequences, such as ACA, resulting in diagonals crossing each other.

		1	2	3	4	5	6	7	8	9	10	11	12	13	14	15
		A	T	G	A	C	A	C	A	G	C	T	T	C	A	G
1	A	X			X		X		X						X	
2	T		X									X	X			
3	G			X						X						X
4	A	X			X		X		X						X	
5	C					X		X			X			X		
6	A	X			X		X		X						X	
7	C					X		X			X			X		
8	A	X			X		X		X						X	
9	A	X			X		X		X						X	
10	C					X		X			X			X		
11	T		X									X	X			
12	T		X									X	X			
13	C					X		X			X			X		
14	A	X			X		X		X						X	
15	G			X						X						X

Figure 3. Dot matrix with all contiguous diagonals of length 3 or longer highlighted.

Apart from the two main diagonals identified in **Figure 3**, the remaining ones are relatively short and may be random occurrences. Looking back at the initial matrix in **Figure 1**, we can see that the multitude of very short diagonals obscures the interpretation and that it results in a cluttered plot. In principle, even a single X, such as the ones in (1,4) or (5,10), represents a "diagonal" of length 1, but such short matches are obviously irrelevant. We therefore need a method to filter out the irrelevant diagonals that are highly likely to represent random "noise", so that a clean and interpretable dot plot can be presented as the result. Ideally, the dot plot should only show matches that have a high likelihood of having biological significance, so that interpretation of the plot can lead to insights into the genetic similarities between the two sequences.

A commonly used method for filtering out irrelevant matches in dot plots is to use a sliding window approach. In short, this means that the algorithm will calculate the degree of similarity between sub-sequences of a given length. It will subsequently include in the plot only the sub-sequences ("diagonals") that have a similarity score above a certain threshold. To illustrate how this is done, we will first go through a simplified toy example to explain the principles. After that we will show how the sliding window approach is used in practice, in a slightly more complex and biologically relevant way, in a tool called Dotmatcher.

To decide which of the matching sub-sequences are relevant to include in the plot, we need to use a scoring scheme. For the initial example, the scoring scheme will be the same very simple one as we used before; it will assign each match between identical nucleotides, for example A in one sequence and A in the other sequence, a score of 1. For pairs of non-identical nucleotides, e.g. A and T, it will assign a score of 0. In addition to the scoring scheme, we need to set a window size, i.e. the length over which the score is calculated for each sub-sequence. In the example shown in **Figure 4**, we are using a window size of 3. For any cell in the matrix, we then calculate the score of the window for which that cell is the midpoint. For this purpose, we redefine the function F to calculate the score for any cell $F(x_i,x_j)$ as the sum of three individual match scores, as follows:

$$F(i,j) = s(x_{i-1},y_{j-1}) + s(x_i,y_j) + s(x_{i+1},y_{j+1}) \qquad Eq\ 3$$

where $s()$ is defined as previously in *Eq* 2. Informally, we can describe this as taking any sub-sequence of length 3 from sequence x, comparing it to any sub-sequence of length 3 from sequence y, and calculating the number of identities between the two sub-sequences. Note that *Eq* 3 can only be applied to windows in one orientation, corresponding to northwest-southeast in the matrix. For completeness, we should also use a similar function to score windows in the opposite orientation, northeast- southwest. However, for clarity and simplicity, we will only include windows oriented in one direction in the following example. In practice, most dot plot tools find matches in both directions, since otherwise inversions would not be identified when genome sequences are compared. Dot plot tools that are primarily used for short sequences, representing a single gene or protein, may have as default setting to do the matching in only one direction. One reason for this is that genes mainly evolve by point mutations, insertions and deletions, and therefore inversions do not need to be considered.

Figure 4 shows an example of how the sliding window method works, using the scoring function from *Eq* 3. We should first note that it is impossible to calculate the scores for the first row and the first column, since when $i = 1$ the value of x_{i-1} is undefined. Differently explained, we cannot calculate the score of a "window" that starts before the beginning of the sequence. Therefore, the first nucleotide of each sequence will not have any window score. Instead, for the first row and column, we have just marked matches with X, as previously. Using the same reasoning, the last row and last column also only contain X for the positions representing identities.

Going over all cells in the matrix we therefore start with the cell (2,2), which is the middle point of the window marked with the letter *a* in **Figure 4**. The window score is 3, because the sequences match in the preceding cell of that diagonal, i.e. $x_1 = y_1$, they match in the central cell of the window, i.e. $x_2 = y_2$, and they match in the subsequent cell, i.e. $x_3 = y_3$.

		1	2	3	4	5	6	7	8	9	10	11	12	13	14	15
		A	T	G	A	C	A	C	A	G	C	T	T	C	A	G
1	A	X		*a*	X		X		X				*b*		X	
2	T		3			1		1	1	1		1	1		1	
3	G			3		1		1		1			1	2		X
4	A	X			3		2		1	1	1		1		1	
5	C		1	1		3		3		1	1			2		
6	A	X			2		3		2	1		1	1		2	
7	C		1	1		3		3		*c* 1	1			2		
8	A	X		1	1	1	2	1	2					1	2	
9	A	X	1		2	1	2	1	1	2			1		1	
10	C		1			2		2		1	2	1		1		
11	T		1				1		1		1	3	1		1	
12	T		1	1	1		*d* 1			1		1	3	1		
13	C			2		2		2			1		1	3		
14	A	X	1		1		2		3			1			3	
15	G			X						X						X

Figure 4. Dot matrix filled in with the sliding window technique, using a window length of 3. The calculation of window scores is illustrated with four examples, *a-d*, which are described and referred to in the text.

For the window with midpoint (11,2), marked with *b* in **Figure 4,** we get a score of 1 since $x_{10} \neq y_1$ (C vs A), $x_{11} = y_2$ (T and T), and $x_{12} \neq y_3$ (T vs G). We can also see how, for the main diagonal, that the window score is 3 across most of the length of that diagonal, but that it drops to 2 for some windows in the vicinity of the A/G substitution. For example, in the window marked with c in **Figure 4,** we get a score of 2 since $x_7 = y_7$ (C in both sequences), $x_8 = y_8$ (A in both), and $x_9 \neq y_9$ (G versus A). An additional example of a window with score 2 is the one with (13,5) as its central cell (*d* in **Figure 4**), since $x_{12} \neq y_4$ (A versus T), $x_{13} = y_5$ (both C), and $x_{14} = y_6$ (both A).

Before proceeding to select the relevant diagonals from the resulting matrix, we should remember that the whole scoring procedure, going over all windows, should also be carried out in the reverse orientation, from the northeast to the southwest corner of the matrix. For the sake of simplicity and clarity, we here refrained from doing that. In practice, most dot plot tools carry out the analysis in both orientations, since sub-sequence matching in the reverse orientations can be biologically relevant, but some tools only do the analysis in the forward orientation, as we do in this example.

As the final step in the procedure, we will locate any score of three in the resulting matrix and then extend that diagonal in both directions for as long as the sequences contain identical nucleotides. For example, starting with the cell (2,2), where the score is 3, we extend the sub-sequence match to (1,1) and to (8,8), despite the score in (8,8) being 2, with the justification that $x_8 = y_8$, meaning that this position should be included in the sub-sequence match. Similarly, starting with the cell (7,5), where the score is 3, we will include the adjacent cells (6,4) and (8,6), thus identifying a matching sub-sequence (ACA) of length 3. The result of this procedure is that two long and three shorter matches are identified, as shown in **Figure 5**.

		1	2	3	4	5	6	7	8	9	10	11	12	13	14	15
		A	T	G	A	C	A	C	A	G	C	T	T	C	A	G
1	A	X														
2	T		3													
3	G			3												
4	A				3		2									
5	C					3		3								
6	A				2		3		2							
7	C					3		3								
8	A						2		2							
9	A															
10	C										2					
11	T											3				
12	T												3			
13	C							2						3		
14	A								3						3	
15	G									X						X

Figure 5. Dot matrix filled in with the sliding window technique, using a window length of 3, retaining only the matches meeting the minimum score criterion. After removal of all low-scoring matches, the matrix is ready to be used as basis for the dot plot visualization.

Now comparing **Figure 1** and **Figure 5**, we can see how the initial matrix based on single nucleotide matches has been substantially cleaned of irrelevant and distracting short matches. Based on this matrix, we can now draw a dot plot that only contains matches that have a reasonable probability of being biologically relevant, rather than just random chance occurrences.

Scoring nucleotide matches with substitution matrices

We will now look at how the sliding window filtering approach is applied in the dot plot tool Dotmatcher. This tool is available at the EBI EMBOSS web server. The EBI (European Institute of Bioinformatics) is part of the EMBL (European Molecular Biology Laboratory), which is a pan-European intergovernmental research organization, with the mission to promote molecular biology research in Europe and develop new technologies. EBI is the bioinformatics branch of this organisation, and one of EBI's many activities is to provide public access to tools for various forms of bioinformatic analysis. As a part of this effort, the EBI maintains EMBOSS, the European Molecular Biology Open Software Suite, which is currently available online through the so-called "Job Dispatcher", which is a unified web-based user interface to the suite of EMBOSS software (Madeira et al., 2024).

Dotmatcher uses essentially the same algorithm as in our previous example, with the main difference that scoring of pairs of nucleotides is done using a so-called substitution matrix (also often called "scoring matrix"). Substitution matrices are very important in bioinformatics since they codify the similarities and differences between nucleotides or amino acids in such a way as to encourage more biologically realistic results in tools for sequence comparison. Basically, substitution matrices are used to adjust for the fact that not all point mutations are equal – some point mutations are more likely than others to affect the properties of the biological molecule that the sequence codes for.

In the case of nucleotides, we can, for example, note that there are two main kinds of point mutations, called transition and transversion. Transition means that a purine is replaced with another purine, i.e. either A (adenine) is replaced with T (thymine) or vice versa. In a transversion, either a purine (A or T) is replaced with a pyrimidine (C or G) or vice versa. Although only 4 of the 12 possible nucleotide substitutions are transitions, it has been established that transitions are generally more frequently observed than transversions. For example, in a study of the substitutional differences between the human and chimpanzee genomes, it was found that 70.3% of point mutations were transitions, corresponding to a 2.4-fold excess of transitions over transversions (Ebersberger et al., 2002). This transition bias has been well known for several decades, and various hypotheses have been proposed for why it exists (Stoltzfus & Norris, 2015). Regardless of its causes, it needs to be accounted for when scoring substitutions in partially matching sub-sequences.

Figure 6a shows parts of a nucleotide substitution matrix called DNAfull (also referred to as EDNAFULL or NUC4.4), which is one of the choices for scoring in Dotmatcher. For matches between identical nucleotides, A-A, C-C, etc., it gives a score of 5, whereas mismatches, such as A-C, A-G, etc, are assigned a score of -4. In addition, this substitution matrix can also be used in cases where sequences include ambiguity symbols, such as S, W, R and Y (**Figure 6b**). These symbols can be used to denote positions in the sequence where variations are observed, for example known SNPs, i.e. single nucleotide polymorphisms. For example, the symbol S can be used for positions where the two alleles C and G have been observed. Accordingly, when A in one sequence is matched with either W (representing nucleotides with weak interactions, i.e. T and A) or R (representing the two purines G and A) in the other sequence, a score of 1 is awarded. The ambiguity symbols are from the standardised IUPAC nomenclature defined by the International Union of Pure and Applied Chemistry ('IUPAC Blue Book', 2022), where the full notation includes additional ambiguity symbols. The substitution matrix shown in **Figure 6a** is therefore only a subset of the complete DNAfull matrix. Note that the matrix is symmetric, since there is no difference in score between symmetrically equal pairs of symbols, such as A-T and T-A.

(a)

	A	T	G	C	S	W	R	Y
A	5	-4	-4	-4	-4	1	1	-4
T	-4	5	-4	-4	-4	1	-4	1
G	-4	-4	5	-4	1	-4	1	-4
C	-4	-4	-4	5	1	-4	-4	1
S	-4	-4	1	1	-1	-4	-2	-2
W	1	1	-4	-4	-4	-1	-2	-2
R	1	-4	1	-4	-2	-2	-1	-4
Y	-4	1	-4	1	-2	-2	-4	-1

(b)

Symbol	*Meaning*	*Mnemonic*
G	guanine	**G**uanine
T	thymine	**T**hymine
A	adenine	**A**denine
C	cytosine	**C**ytosine
S	G+C	**S**trong interactions
W	T+A	**W**eak interactions
R	G+A	pu**R**ine
Y	T+C	p**Y**rimidine

Figure 6. Example substitution matrix with scores to use when matching nucleotide sequences. (a) Parts of the DNAfull substitution matrix, showing eight of the total 15 symbols included in the complete matrix. (b) Interpretation of symbols representing nucleotides (G,T,A,C) or groups of nucleotides (S,W,R,Y) according to the IUPAC notation.

We will now create a dot plot for our example sequences in the Dotmatcher tool[1]. **Figure 7** shows the web interface to Dotmatcher. We first chose DNA as sequence type, and then pasted in the two sequences, using a header line for each, such as ">seq1", so that the input

[1] At the time of writing, the Dotmatcher tool was available at the first address shown below. If it has moved at the time of reading, then it is likely that it can be found by searches starting at the Job Dispatcher main page or at the EBI main page.
www.ebi.ac.uk/jdispatcher/seqstats/emboss_dotmatcher
www.ebi.ac.uk/jdispatcher
www.ebi.ac.uk

is in FASTA[2] format. Dotmatcher does not strictly require the input to be in this format, so just pasting in the nucleotide sequences will also work. In that case, the tool will itself add a dummy header line with a made-up identifier, such as:

>EMBOSS_001

Even though not required by this tool, it is a good habit to use FASTA format whenever it is accepted, since it is widely used and considered the *de facto* standard text format for nucleotide and amino acid sequences. Also, in this case, Dotmatcher will show the identifiers ("seq1" and "seq2" in the example) as labels for the axes of the dot plot. Therefore, providing FASTA header lines containing sequence identifiers is a convenient way of documenting the plots made. In the "Parameters" section, we have chosen DNAfull as substitution matrix, a window size of 3 and a score threshold of 15. Considering the match score for nucleotide identities in DNAfull being 5, this corresponds to requiring all three positions in the window to be identities.

[2] en.wikipedia.org/wiki/FASTA_format

Figure 7. User interface to the Dotmatcher tool at the EBI-EMBL Job Dispatcher web server (screenshot from March 2025). To run the example described in the text, we need to choose “DNA” as sequence type, paste in the two sequences, then set the three parameters “Matrix”, “Window size” and “Threshold” accordingly.

With these settings we should expect to identify a similar set of sub-sequence matches as the ones we found in **Figure 5** using the simple scoring scheme. Looking at the Dotmatcher results in **Figure 8**, it may at first glance seem that the results differ. However, we need to consider that the plots drawn by Dotmatcher show the lower left corner as origin, whereas we set the origin at the upper left corner in **Figure 5**. It is here worth considering that different dot plot tools set the origin differently. To see that the two results are identical, we could rotate one of the plots by 90 degrees.

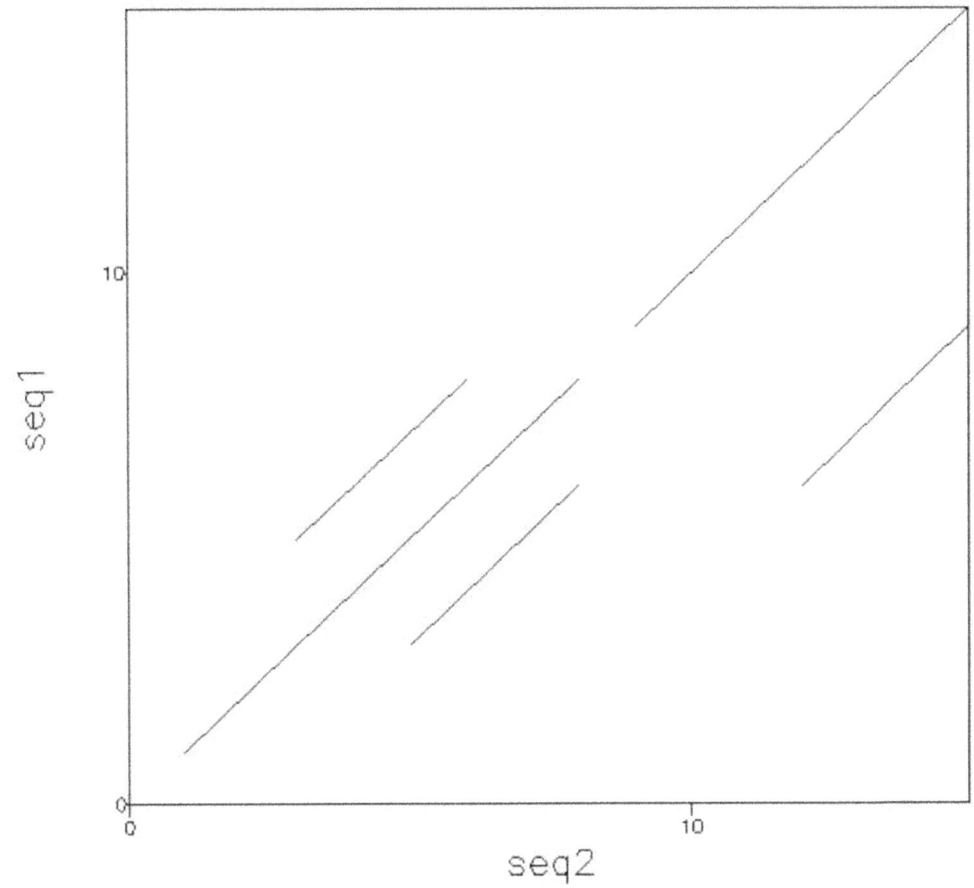

Figure 8. Dotmatcher results for the example input shown in **Figure 7**.

Using dot plots to compare homologous genes

We will now consider a more biologically realistic example, where the mRNA sequences of two related genes are compared using Dotmatcher. The example gene, named *S100A3*, codes for the S100 calcium-binding protein A3. We will be comparing mRNA sequences representing transcripts of this gene from human (*Homo sapiens*) and rat (*Rattus norvegicus*). For these relatively closely related organisms, we would expect the similarities between the sequences to be clearly visible in a dot plot, but we would also expect to see some differences.

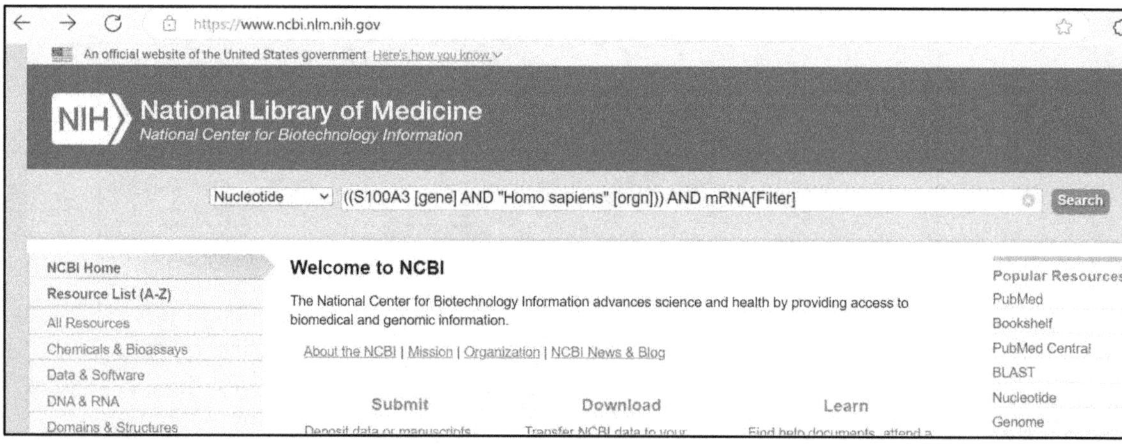

Figure 9. The two mRNA sequences for S100A3 genes were found in the Genbank nucleotide database using queries as shown in this screenshot.

We start by searching for the two sequences in the Genbank[3] nucleotide sequence database (Benson et al., 2013). Genbank is one of the most important publicly available sources for nucleotide sequence data, together with the DNA Data Bank of Japan and the European Nucleotide Archive, which are the three partners in the International Nucleotide Sequence Database Collaboration (Arita et al., 2021). One way of finding entries in Genbank that contain mRNA sequences from the human and rat S100A3 is to visit the main web page of NCBI (National Center for Biotechnology Information), as shown in **Figure 9**, choose "Nucleotide" in the menu "All Databases", and type in search queries as follows, one at a time:

```
((S100A3 [gene] AND "Homo sapiens" [orgn])) AND mRNA[Filter]
((S100A3 [gene] AND "Rattus norvegicus" [orgn])) AND mRNA[Filter]
```

At the time of writing (March 2025), these queries gave four and eleven "hits", i.e. matching database entries. From these entries, we chose the one with accession number NM_002960

[3] www.ncbi.nlm.nih.gov/genbank

from the list of human hits and NM_053681 from the list of rat entries. To reproduce the Dotmatcher example shown in the following, you may either repeat the searches as shown in **Figure 9** or just type in the accession numbers in the Genbank search field to find each entry directly.

To download the sequences from Genbank, there are two main methods, illustrated in **Figure 10**. We can follow the link "FASTA" to see the sequence in FASTA format in the browser, then copy and paste it into the Dotmatcher tool. Alternatively, we can follow the "Send to" link to open a popup window with download settings. When choosing "File" in the section "Choose Destination" and "FASTA" in the "Format" menu, your browser should start downloading a text file containing the FASTA entry for the sequence. When you have saved this text file, it should then be possible to upload it to Dotmatcher.

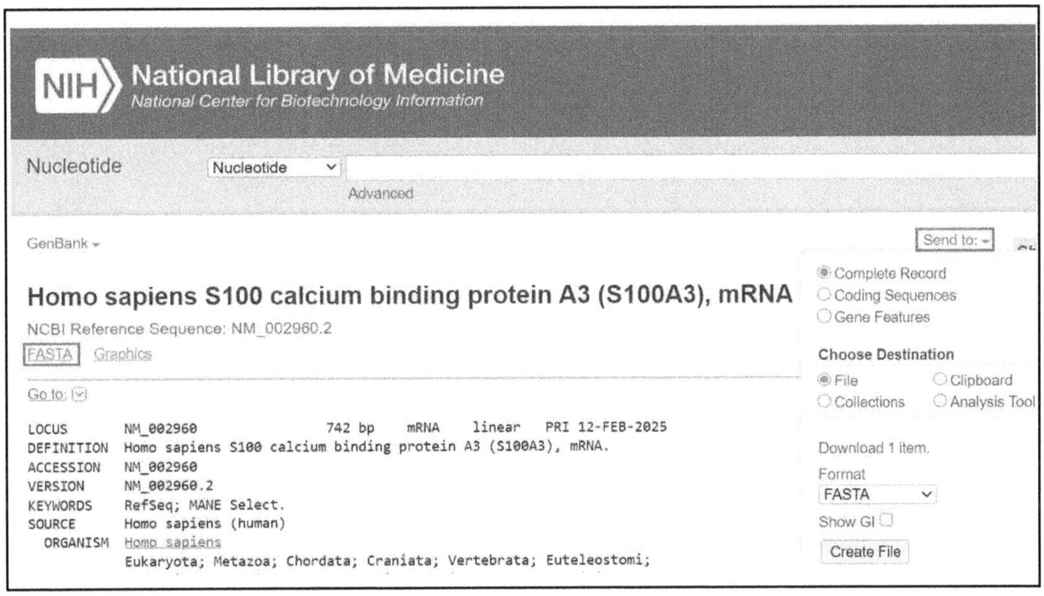

Figure 10. Two methods for obtaining FASTA format sequences from Genbank entries, indicated by the read boxes.

When running Dotmatcher with the DNAfull substitution matrix, the default setting for window size is 10 and the default score threshold is 23. Using these default settings, we obtained the result shown in **Figure 11**. In this plot, the human sequence is shown vertically and the rat sequence horizontally. On one hand, it is possible to discern a long stretch of diagonal starting near the southwest corner and stretching across large parts of the plot, but on the other hand, the interpretation is obscured by a large number of short matches.

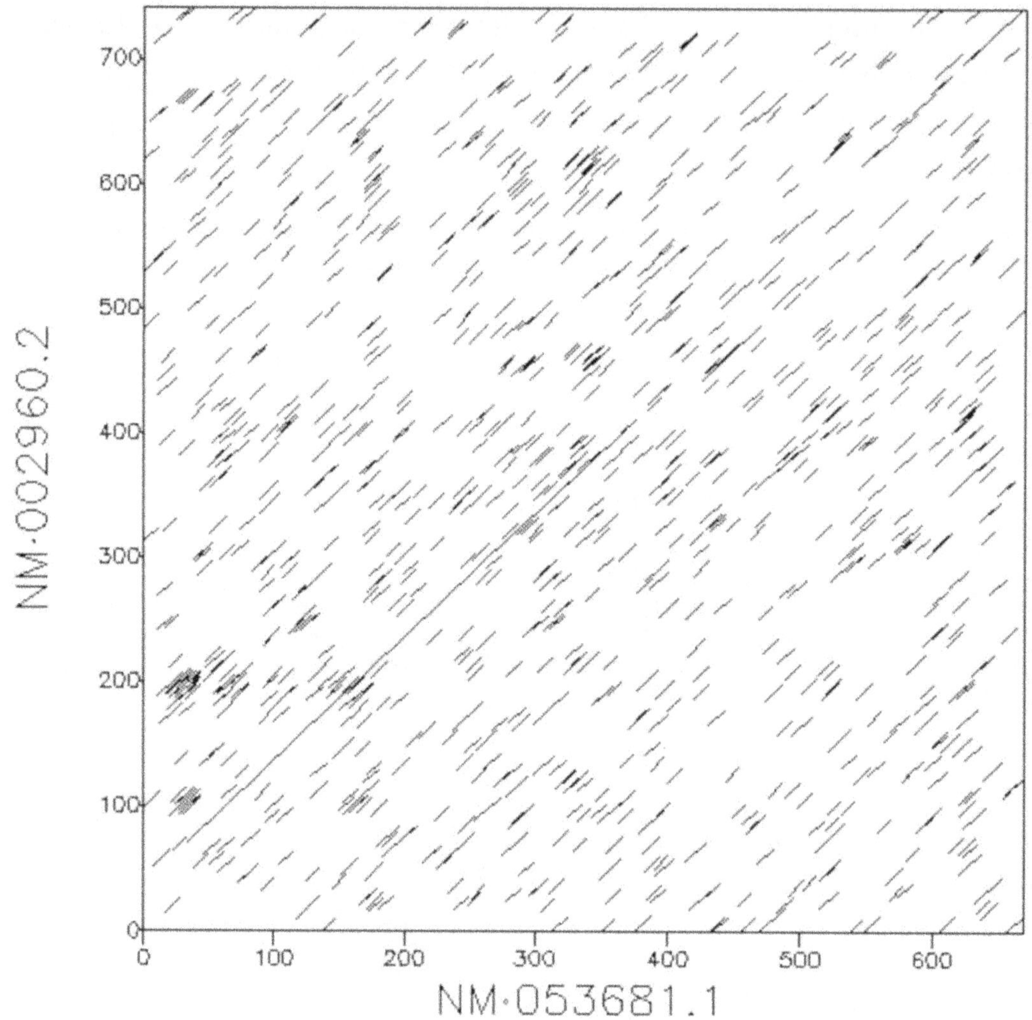

Figure 11. Dot plot of mRNA sequences from the human (vertical axis) and rat (horizontal axis) S100A3 gene, encoding the S100 calcium-binding A3 protein. Generated with Dotmatcher, using default settings, i.e. substitution matrix = DNAfull, window size = 10 and score threshold = 23.

The most straightforward way of trying to "clean up" the plot is by filtering out more of the irrelevant matches by increasing the score threshold. Therefore, using the same input sequences, the same substitution matrix, and the same window length, we can try different settings of the score threshold to achieve this filtering. When setting the threshold at 40 (**Figure 12**, left), we get a much clearer plot, where most of the noise has been removed while the main diagonal remains. We can now clearly see that the sequences are similar across approximately 300 positions, starting near the coordinates (100, 100) and ending in the vicinity of (400, 400). After that, there is some detectable similarity, but at a lower level, since the diagonal becomes broken into smaller pieces, although these are lined up in an almost straight line all the way to the northeast corner. Altogether, this indicates that the two mRNA sequences are highly similar overall. The plot to the right in **Figure 12** shows what happens when we overdo the filtering by choosing a too strict score threshold, in this case 50. For a window size of 10, the maximum window score that can be reached with the DNAfull scoring matrix is 50. Therefore, this score is achieved only when the two sub-sequences in the window are identical in all positions. With these strict settings, some of the relevant findings are lost since the matching segments close to the northeast corner are no longer seen in the plot.

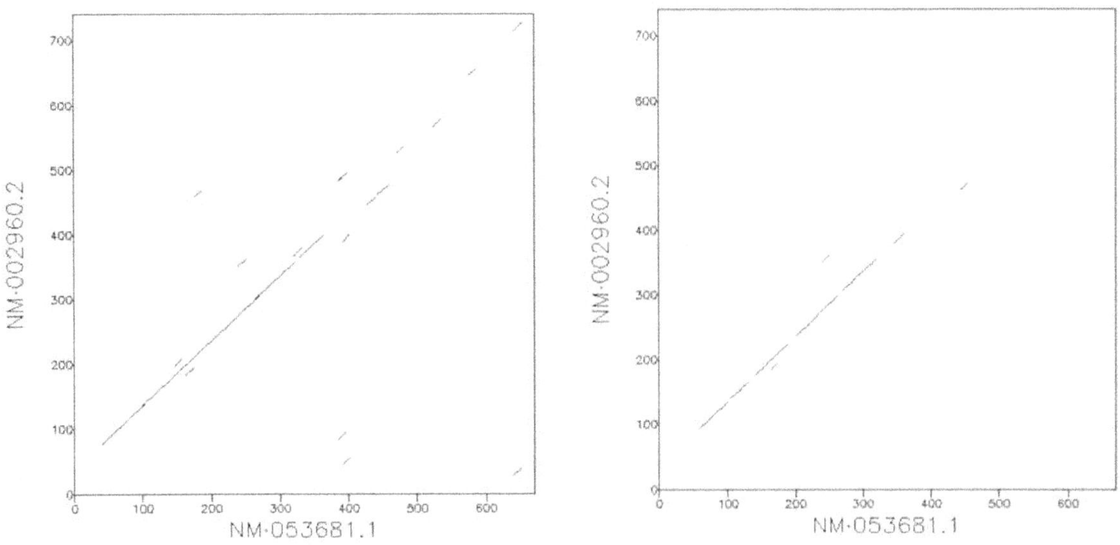

Figure 12. Dotmatcher plots using the same input sequences as in **Figure 11** but with stricter settings for the score threshold. Left: score threshold = 40. Right: score threshold = 50.

As this example illustrates, using a dot plot tool in practice requires setting the parameters in a balanced way, so that as much of the relevant results are retained while simultaneously filtering out as much as possible of the noise. The example also illustrates that a tool's default settings, even if carefully chosen based on experience and expertise, will not always produce the best results. The reason for this is that there is so much variation in the data and in the underlying biological factors that we aim to discover when using the tool, that it is a near impossible task to find default parameter settings that will produce the best results in all cases.

Using dot plots to compare homologous protein sequences

As we have already seen, dot plots can be used to obtain a rough estimate of the evolutionary distance between pairs of sequences and to see which parts of the sequences match with each other. We will here apply this technique to protein sequences, which requires a more complex scoring function than the one used for nucleotide sequences. Whereas there are only four nucleotides, there is a total of 20 commonly occurring amino acids. In addition, amino acids can be characterised by several types of physical and chemical properties, which in turn affect the biochemical properties of the protein molecule. How substitution matrices are generated from protein sequence data and how they are used in various methods for protein sequence comparisons is therefore a topic that requires lengthy explanation and detailed study. In this section, we will just look at one example substitution matrix for amino acids and show how it is used in the Dotmatcher tool.

	C	S	T	P	A	G	N	D	E	Q	H	R	K	M	I	L	V	F	Y	W
C	9																			
S	-1	4																		
T	-1	1	5																	
P	-3	-1	-1	7																
A	0	1	0	-1	4															
G	-3	0	-2	-2	0	6														
N	-3	1	0	-2	-2	0	6													
D	-3	0	-1	-1	-2	-1	1	6												
E	-4	0	-1	-1	-1	-2	0	2	5											
Q	-3	0	-1	-1	-1	-2	0	0	2	5										
H	-3	-1	-2	-2	-2	-2	1	-1	0	0	8									
R	-3	-1	-1	-2	-1	-2	0	-2	0	1	0	5								
K	-3	0	-1	-1	-1	-2	0	-1	1	1	-1	2	5							
M	-1	-1	-1	-2	-1	-3	-2	-3	-2	0	-2	-1	-1	5						
I	-1	-2	-1	-3	-1	-4	-3	-3	-3	-3	-3	-3	-3	1	4					
L	-1	-2	-1	-3	-1	-4	-3	-4	-3	-2	-3	-2	-2	2	2	4				
V	-1	-2	0	-2	0	-3	-3	-3	-2	-2	-3	-3	-2	1	3	1	4			
F	-2	-2	-2	-4	-2	-3	-3	-3	-3	-3	-1	-3	-3	0	0	0	-1	6		
Y	-2	-2	-2	-3	-2	-3	-2	-3	-2	-1	2	-2	-2	-1	-1	-1	-1	3	7	
W	-2	-3	-2	-4	-3	-2	-4	-4	-3	-2	-2	-3	-3	-1	-3	-2	-3	1	2	11

Figure 13. The BLOSUM62 substitution matrix.

The scoring matrix used here is called Blosum62 and is shown in **Figure 13**. A few main properties of this matrix should be noted, without going into details. First, we can see that all identity matches give positive scores, as expected, but that some identities are scored higher than others. The highest score (+11) is awarded for matching tryptophan (W) with itself and the second highest (+9) for matching cysteine (C) with itself. This can be explained in a simplified way by noting that each of these amino acids has important and relatively unique properties. Tryptophan occurs quite rarely in proteins, but when it does occur, it tends to play important roles, for example in the anchoring of membrane proteins within the cell membrane (Khemaissa et al., 2021). Cysteine is also one of the least abundant amino acids, but is often observed in sites that are functionally important, such as regulatory, catalytic, or binding sites (Marino & Gladyshev, 2010). Pairs of cysteines often form disulfide bonds, which play an important role in the structural stability of the protein molecule (Fass, 2012). The consequence of this in terms of protein sequence comparison, is that if both sequences contain a W, or both contain a C, in a matching sub-sequence, this is a strong indication of relatedness between the two proteins.

Further, when observing the scores in the Blosum62 matrix, we can see some groups of amino acids that are similar to each other and share some key properties. For example, the MILV group (methionine, isoleucine, leucine, and valine) are almost freely interchangeable, since any substitution within this group receives a positive score. This is because these amino acids are all hydrophobic, and it tends to be of little consequence for the structure or function of the protein if one is replaced by another. Thus, in a scoring function for generating dot plots between protein sequences, different pairs of matches and mismatches will be treated rather differently, in a way that reflects the properties and roles of the amino acids involved in the match or mismatch.

The description may give the impression that the scores in Blosum62 were decided by analysis of amino acid properties. However, to avoid the introduction of human bias and possible effects of incorrect assumptions, the matrix was derived from observations of substitution frequencies in aligned sequence segments from groups of related proteins. The procedure is described in the article that defined the BLOSUM (BLOcks SUbstitution Matrix) method (Henikoff & Henikoff, 1992). In very brief summary, the authors collected about 2000 "blocks", which are short alignments representing conserved regions in protein families. The blocks came from about 500 different protein families, to have a wide representation of structural and functional properties. By analysing each column in these blocks, the relative frequencies of amino acids and their substitutions were counted. By comparing with the substitution frequencies expected by chance, they also calculated the log odds ratio for each of the possible 20 + 19 + ... + 1 = 210 substitution pairs for the 20 standard amino acids. The log odds ratio was calculated in bit units, multiplied by a scaling factor of 2, and rounded to the nearest integer. Following this, scores > 0 represent frequencies that are higher than expected by chance, while scores < 0 represent frequencies that are lower than expected by chance.

The researchers who designed the BLOSUM method used blocks with different percent identity thresholds (Henikoff & Henikoff, 1992). The Blosum62 matrix, for instance, was derived from blocks where the sequences had pairwise average percent identity of up to

62%. The reason given by the authors for setting an upper limit of sequence identity was to avoid bias from groups of near-identical sequences and improve sensitivity for detection of distantly related sequences. Many tools offer a selection of different matrices, such as Blosum30 and Blosum45 for detection of weakly similar sequences, or Blosum80 for comparison of highly similar sequences. Experimentation done by the original authors indicated that Blosum62 provided the best overall performance (Henikoff & Henikoff, 1992). It was thereafter adopted as the default scoring matrix for well-known tools such as the BLAST homology search algorithm (Altschul et al., 1990) and is frequently recommended as one of the best default choices for sequence similarity search programs in general (Pearson, 2013).

To illustrate the use of amino acid substitution matrices in Dotmatcher, we will use myoglobin protein sequences from different organisms. Myoglobin binds to oxygen and is found in cardiac and skeletal muscle tissue, where it plays an important role in the storage and transportation of oxygen (Ordway & Garry, 2004). In vertebrates, the gene duplication event leading to the appearance of myoglobin seems to have taken place around 500 million years ago (Hardison, 2012), and it is therefore possible to study the evolution of the myoglobin protein across a wide range of vertebrate species. For the following dot plot examples, we have selected myoglobin protein sequences from human, the crab-eating macaque (*Macaca fascicularis*), rat (*Rattus norvegicus*) and the loggerhead sea turtle (*Caretta caretta*). For each plot, the human myoglobin sequence will be compared with one of the other sequences. The evolutionary distances between sequences in these plots will then range from approximately 29 million years (human versus macaque), through 87 million years (human versus rat), to 319 million years (human versus sea turtle), with approximate distances estimated by the TimeTree[4] (Kumar et al., 2022).

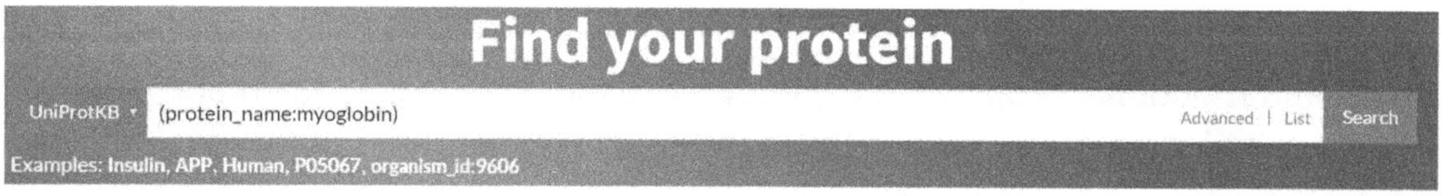

Figure 14. Search query to find entries with myoglobin as protein name in the UniProt database.

Sequences for this example were downloaded from the UniProt[5] protein sequence database (The UniProt Consortium, 2025). The sequences can be found by searching for entries with "myoglobin" as protein name, as in the query shown below and in **Figure 14**, and then browsing through the hits to locate the relevant entries:

(protein_name:myoglobin)

---

[4] timetree.org

[5] www.uniprot.org

Alternatively, we can search for the entry names or accession numbers shown in **Table 1**.

Table 1. UniProt entries used in dot plot protein sequence comparisons.

Entry name	Accession number	Organism
MYG_HUMAN	P02144	*Homo sapiens*
MYG_MACFA	P02150	*Macaca fascicularis*
MYG_RAT	Q9QZ76	*Rattus norvegicus*
MYG_CARCR	P56208	*Caretta caretta*

From the download pop-up window in UniProt (**Figure 15**), the selected sequences can either be saved to a text file or copy-pasted from the preview box into the Dotmatcher web interface.

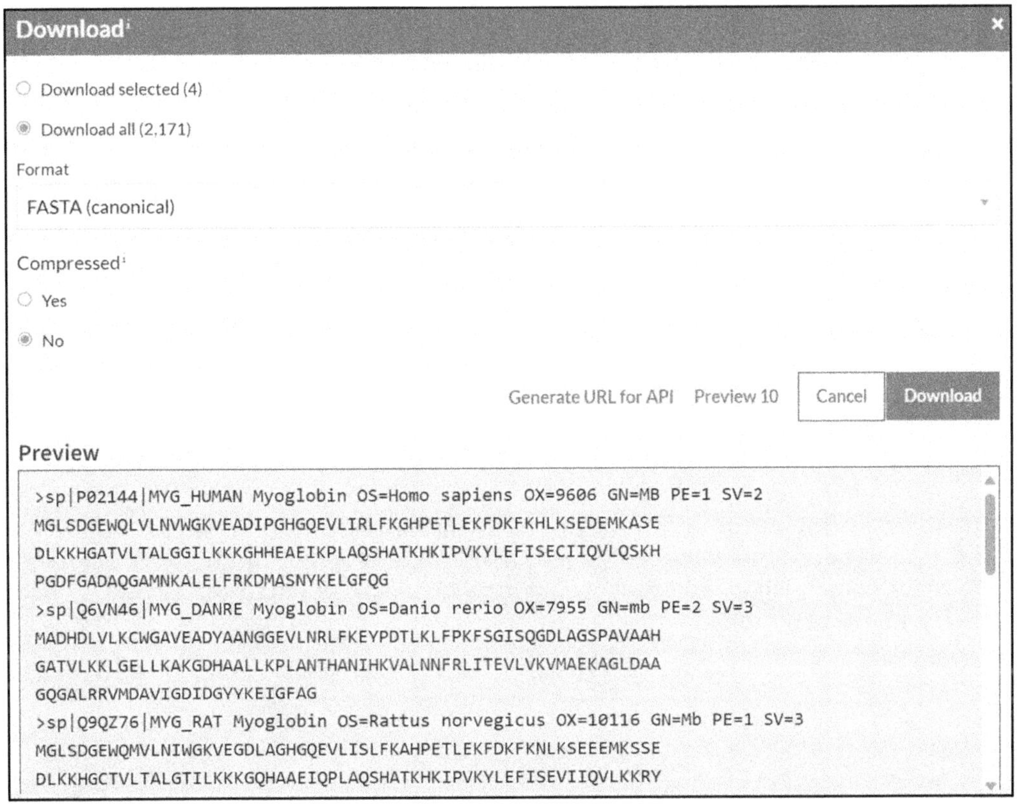

Figure 15. Downloading selected sequences from the UniProt protein sequence database.

Dot plots for human versus macaque, rat and sea turtle were generated with Dotmatcher using "Protein" as sequence type, choosing Blosum62 as substitution matrix, a window size of 10 and setting the score threshold to 50 (**Figure 16**). We should note here that these settings translate to requiring sub-sequences of length 10 to have a minimum average score of 5 per position, which, considering the values in the Blosum62 matrix, is a relatively strict requirement. Some identity matches, such as isoleucine matched with itself, give scores lower than 5, and any mismatch within the window would have to be compensated for by one or more identity matches of rather unique and high-scoring amino acids. Despite these strict requirements, substantial similarity is seen in all three comparisons (**Figure 16**).

As the series of dot plots in **Figure 16** show, there is a clear correlation between how closely or distantly related the organisms are and the degree of similarity between the protein sequences. When the human protein is compared to that from another primate, the crab-eating macaque, the main diagonal is almost complete. When instead compared to the homologous protein from a rodent, i.e. rat, the diagonal is broken up into four segments, which together cover only about half of the distance from corner to corner. Finally, when the human sequence is compared to one from a turtle, i.e. a reptile, only one short match appears. Considering that humans and turtles are separated by over 300 million years of evolution, it may, however, be considered quite remarkable that significant similarity can still be observed, and that we can find it with such a simple method as a dot plot.

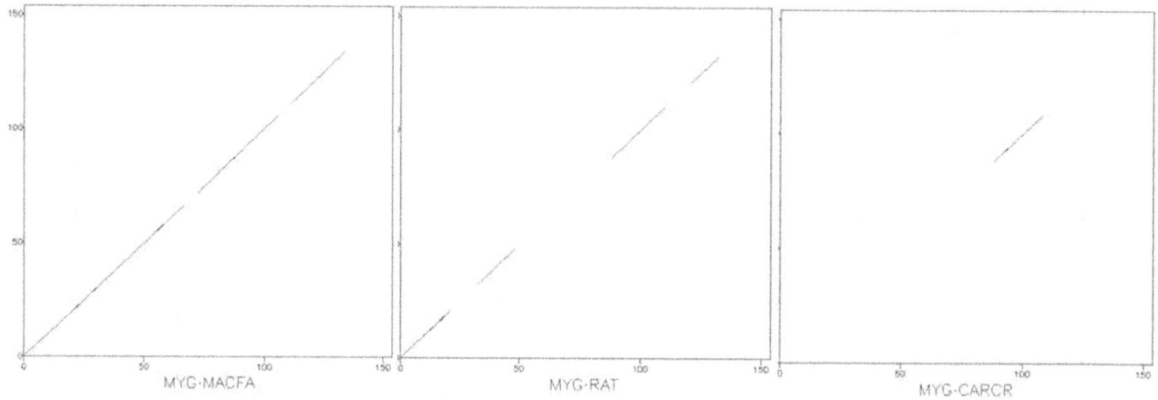

Figure 16. Dotmatcher dot plots of myoglobin protein sequences, comparing human (MYG_HUMAN) with, from left to right, the crab-eating macaque (MYG_MACFA), rat (MYG_RAT) and the loggerhead sea turtle (MYG_CARCR). These comparisons represent evolutionary distances of approximately 29, 87 and 319 million years, respectively (Kumar et al., 2022).

Plotting a protein sequence against itself to find repeats and regions with compositional bias

Dot plots are sometimes used to discover properties of a sequence without comparing it to any other sequence. One feature that is easily visualised when a sequence is plotted against itself is the presence of repeat regions. As the first example of this, we will here study the sequence of an ice nucleation protein. Some bacteria possess such proteins, which enable them to initiate crystallisation of water at relatively high temperatures, thereby inducing frost damage to plants and obtaining nutrients from them (Roeters et al., 2021). Ice nucleation proteins are located at the outer surface of the membrane of these bacteria, and it is believed that they produce templates for the assembly of seed crystals of ice. They also contain a repetitive primary structure that seems to be responsible for aligning water molecules to the seed crystal during the ice formation process (Wolber & Warren, 1989).

According to the PROSITE (Sigrist et al., 2013) documentation entry[6] for the Bacterial ice-nucleation proteins octamer repeat, the repeat region takes up around 80% of the total length of ice-nucleation protein sequences. The repeat region contains a conserved 48 residues long repeat, which itself consists of three blocks of 16 residues, where the first eight residues of each block are identical.

Looking at a specific ice nucleation protein, ice nucleation protein InaA from the gram-negative bacterium *Pantoea ananas*, we can see that the repeat region corresponds to a repeated secondary structure in the predicted protein structure (**Figure 17**). The molecule, as predicted by AlphaFold (Jumper et al., 2021), has a long protruding substructure, visually similar to a ladder, consisting of beta strands. This ladder-like structure is likely involved in aligning water molecules to the seed ice crystal.

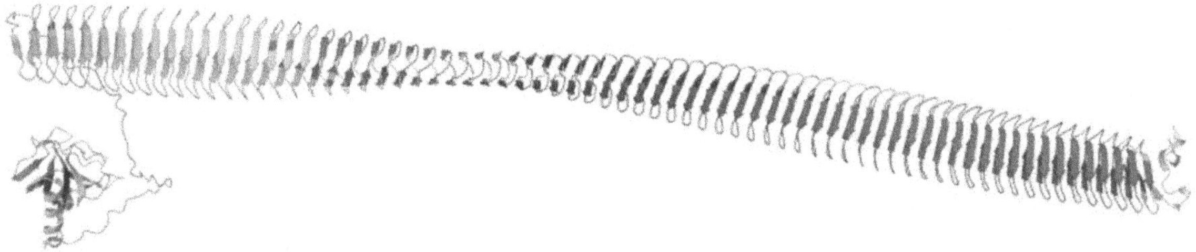

Figure 17. Predicted 3D structure of the Ice nucleation protein InaA from the bacterium *Pantoea ananas*, UniProt sequence entry ICEA_PANAN (P20469). Structure visualization from the corresponding AlphaFold database entry.

[6] prosite.expasy.org/PDOC00283

The amino acid sequence of UniProt entry ICEA_PANAN (accession number P20469) was downloaded from UniProt and analysed in preparation for the dot plot example. Using the tool ScanProsite[7] (de Castro et al., 2006) to search for occurrences of the Prosite pattern for Bacterial ice-nucleation proteins octamer repeat revealed 49 occurrences throughout the ICE_PANAN sequence, with the first occurrence starting at residue number 178 and the last ending at position 1081. As seen in **Figure 18**, the pattern is A-G-Y-G-S-T-x-T, which will match any sub-sequence of length eight that contains the specified amino acids in positions 1-6 and 8, whereas the x at position 7 represents that any amino acid can occur in that location.

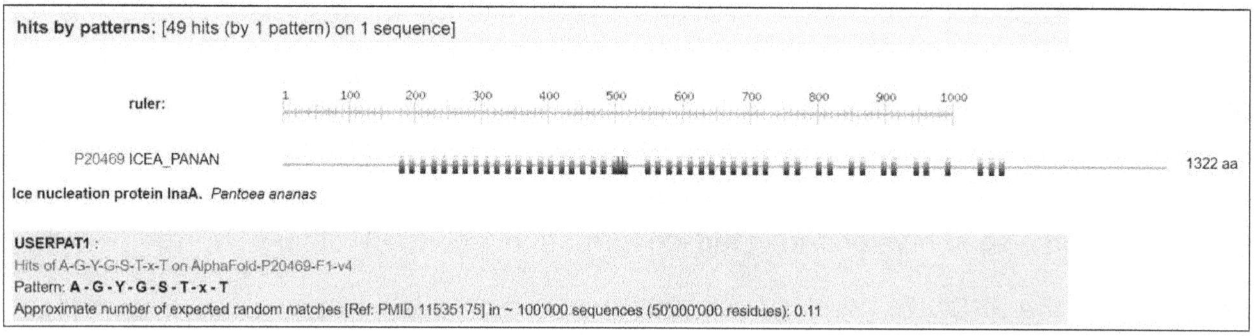

Figure 18. Occurrences of the Bacterial ice-nucleation proteins octamer repeat in the ICEA_PANAN sequence, according to ScanProsite search results matching the pattern AGYGSTxT to the protein sequence.

In **Figure 19** we see the results of plotting the ICE_PANAN amino acid sequence against itself. The first approximately 170 positions of the sequence do not contain any repeats. From that position onwards, the very large number of repeats leads to a "cloud", approximately square-shaped, of matches, clearly showing the beginning and end of the repeat region.

[7] prosite.expasy.org/scanprosite/

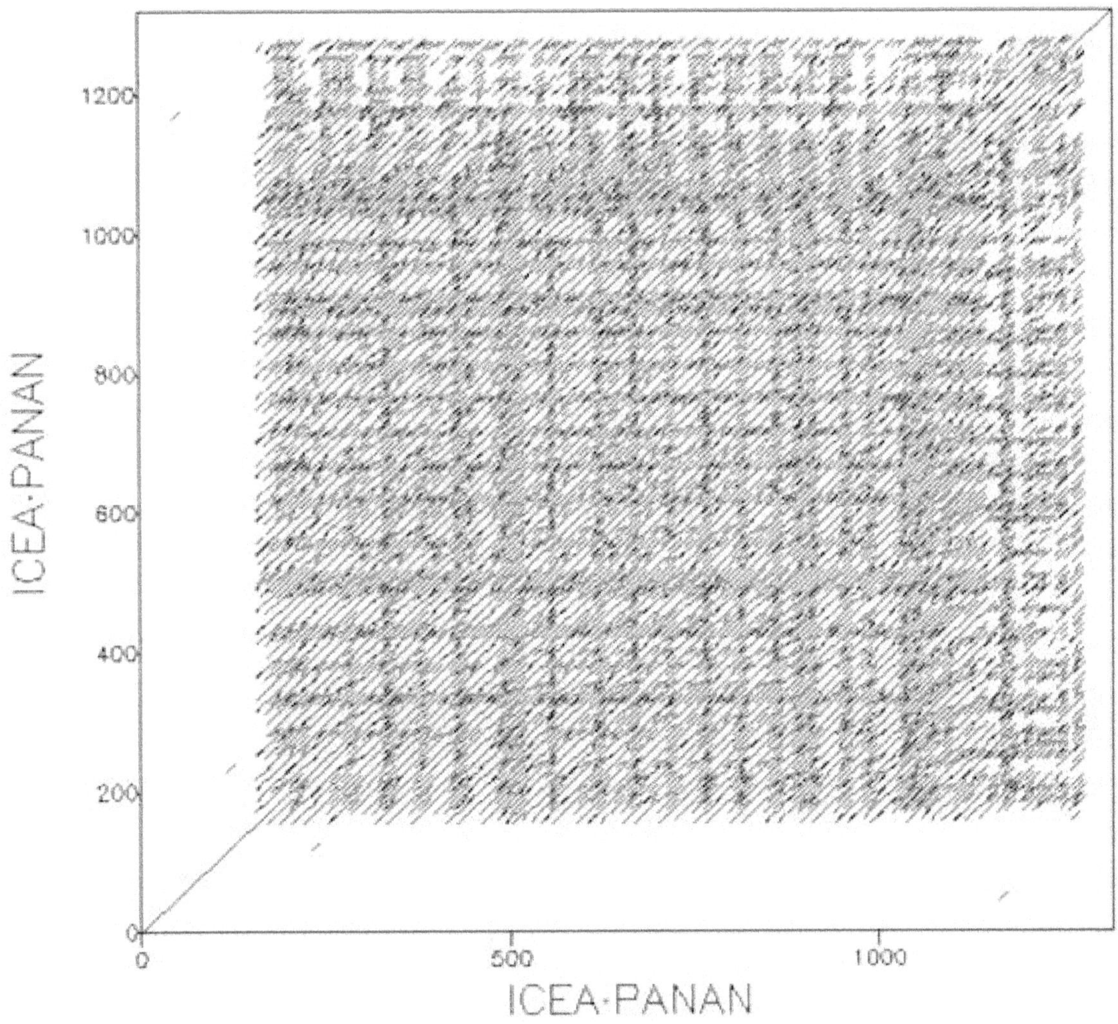

Figure 19. Dotmatcher plot of the repeat-enriched ICEA_PANAN versus itself, generated with default parameter settings.

As a second example of identification of repeat regions using dot plots, we will study a protein named Eukaryotic translation initiation factor 3 subunit A (eIF3A). This protein is the RNA-binding component of the translation initiation factor 3 complex, which is required for the initiation of protein synthesis (Masutani et al., 2007). The human sequence of eIF3A and description of its properties can be found in the UniProt entry EIF3A_HUMAN, with accession number Q14152. The dot plot for EIF3A_HUMAN, shown in **Figure 20**, indicates a distinct repeat region, the extent of which is represented by the bar marked with *a*.

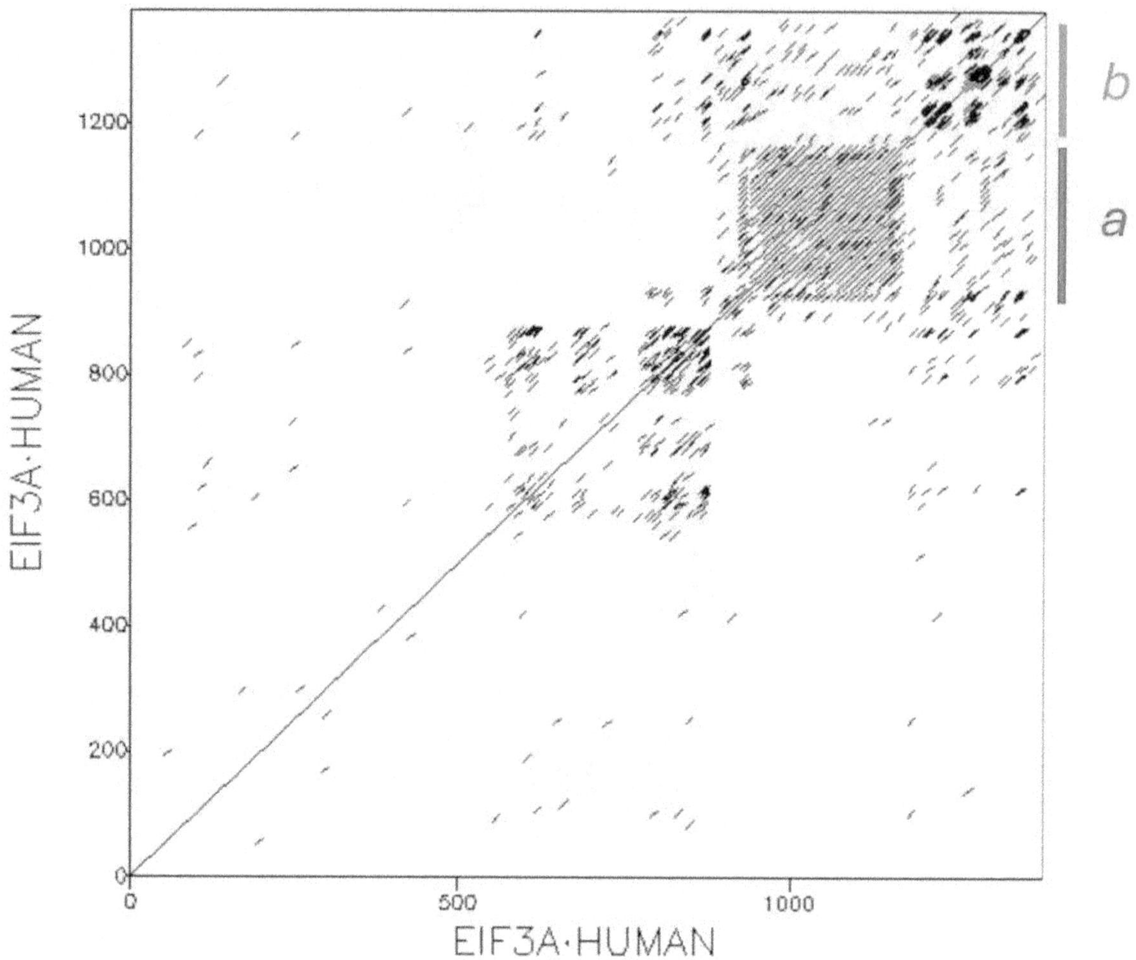

Figure 20. Dotmatcher plot of the EIF3A_HUMAN protein sequence versus itself, generated with default parameter settings. The bar (*a*) indicates a region containing 25 approximate repeats. The bar (*b*) indicates a region with compositional bias.

Looking at the Q14152 entry in UniProt, it states that the region starting at position 925 in the sequence and ending at 1172 contains 25 copies of a ten amino acids long repeat sequence (**Figure 21**). The repeat is approximate, meaning that the instances of the repeat are similar to each other, rather than being exactly identical. The UniProt entry shows the approximate repeat sequence in the format of a Prosite pattern (Sigrist et al., 2013), which specifies the amino acids that occur in each position. The pattern in this case is the following:

[DE]-[DE]-[DE]-R-[SEVGFPILV]-[HPSN]-[RSW]-[RL]-[DRGTIHN]-[EPMANLGDT]

The pattern should be interpreted as specifying that the repeat sequence begins with either D or E (the first [DE] element), it then has either D or E also in the second and third positions, followed by R in the fourth position. In the fifth position, it has any of the amino acids shown in the fifth element of the pattern, i.e. S, E, V, G, F, P, I, L or V, as specified by [SEVGFPILV]. Continuing in the same manner, every instance of the repeat sequence should match all ten elements of the pattern.

+	Repeat	925-934	1
+	Region	925-1172	25 X 10 AA approximate tandem repeats of [DE]-[DE]-[DE]-R-[SEVGFPILV]-[HPSN]-[RSW]-[RL]-[DRGTIHN]-[EPMANLGDT]
+	Repeat	935-942	2; truncated
+	Repeat	943-952	3
-	Repeat	953-962	4
	Sequence: DDDRVPRRGM		
-	Repeat	963-972	5
	Sequence: DDDRGPRRGP		

Figure 21. Information in the Q14152 entry in UniProt states that a repeat region starting at position 925 and ending at 1172 contains 25 copies of an approximate repeat sequence of ten amino acids. The repeats show variation as indicated by the pattern, starting with [DE], representing that the first amino acid in the repeat is either D or E. The sequences of repeats number 4 and 5 are shown as examples.

As an example, we see in **Figure 21** that repeat sequence number 4 is DDDRVPRRGM, which matches the pattern as follows, with amino acids in bold font indicating how the sequence matches the pattern in each of the ten elements:

[**D**E]-[**D**E]-[**D**E]-**R**-[SEVGFPIL**V**]-[H**P**SN]-[**R**SW]-[**R**L]-[DR**G**TIHN]-[EP**M**ANLGDT]

To see that these repeats, although not identical to each other, generate matches in the dot plot, we show below an alignment of repeat sequences number 4 and 5. When the plot was generated, with default parameter settings, the Blosum62 substitution matrix was used, and the score threshold for windows of length ten was 23. In the alignment we compare the two repeat sequences position by position and show the score below each amino acid pair. For instance, both sequences begin with D, which gives a score of 6 (according to Blosum62, see **Figure 13**). In the fourth position both sequences contain R, which, when matched with itself, gives a score of 5. In the sixth position, matching V with G gives a score of -3, which is indicated with bold font. When having set the scores for all positions, summing up the ten individual scores gives a window score of 41, which is well above the threshold of 23.

DDDRVPRRGM
DDDRGPRRGP
6665**3**7556**2** -> Window score: 6+6+6+5**-3**+7+5+5+6**-2** = 41

Thus, we can assume that the presence of 25 similar, but not identical, repeat sequences will result in a cluster of short matches in the self-comparison of the full protein sequences, as was indicated in **Figure 20**.

There are other parts of the plot that may seem to also represent repeats. For example, the region near the end of the sequence, indicated by the bar marked with the letter *b* in **Figure 20**. However, when looking again at the information in the UniProt entry, it is stated that the region 1177-1328 has compositional bias towards basic and acidic residues. To confirm this, we copied the corresponding part of the sequence and used the tool ProtParam[8] (Gasteiger et al., 2005) to calculate its amino acid composition. This revealed that this region has a very skewed distribution, where three amino acids together make up two thirds (66.5%) of the residues: arginine (R) 30.3%, asparagine (D) 19.1% and glutamic acid (E) 17.1%.

This bias in itself will result in many random occurrences of short sub-sequence matches which are not due to actual repeats. It can rather be contributed to the fact that the scoring parameters of the tool (window length 10 and score threshold 23) are set under the assumption that the frequencies of individual amino acids do not deviate significantly from the background distribution. When we have very strong compositional bias, these assumptions do not hold. To see this intuitively, consider a case where, for example, three of the amino acids together would make up 100% of the symbols in the sequence. We would then in practice have a sequence in a three-letter alphabet and window scores being evaluated with a threshold tuned for 20-letter alphabet sequences. Or, for an even more extreme example, if a single symbol was seen in every position of the sequence, then obviously every sub-sequence would match perfectly with every other sub-sequence.

As a conclusion from the previous example, we see that it can be difficult to distinguish repeat regions from regions with compositional bias by only looking at a dot plot. As a general rule of thumb, however, repeats tend to appear in a more regular pattern, whereas the sub-sequence matches in regions with compositional bias tend to be a little noisier and more random. Looking again at the regions marked *a* and *b* in **Figure 20**, the first appears more orderly and "organised", with a regular repeat pattern, whereas the second has more randomly spread-out sub-clusters of matches, corresponding to sub-sequences that happen to match due to the compositional bias. Making this distinction solely based on visual inspection of the plot is, of course, not recommended. The visual patterns should be considered as a basis for hypotheses to be evaluated, confirmed or rejected, with the use of other tools. The central part of the plot in **Figure 20**, approximately between positions 600

[8] web.expasy.org/protparam/

and 850, presents another case which can be investigated by the reader as an exercise in dot plot interpretation.

As a final example in this chapter, we will study the human insulin receptor protein, INSR_HUMAN (P06213). Looking first at the result of plotting this sequence against itself, it seems overall that there are only a few scattered short matches that appear to be random occurrences (**Figure 22**). The exception to this is a region, marked with the rectangle, which seems to have a slightly higher density of matches.

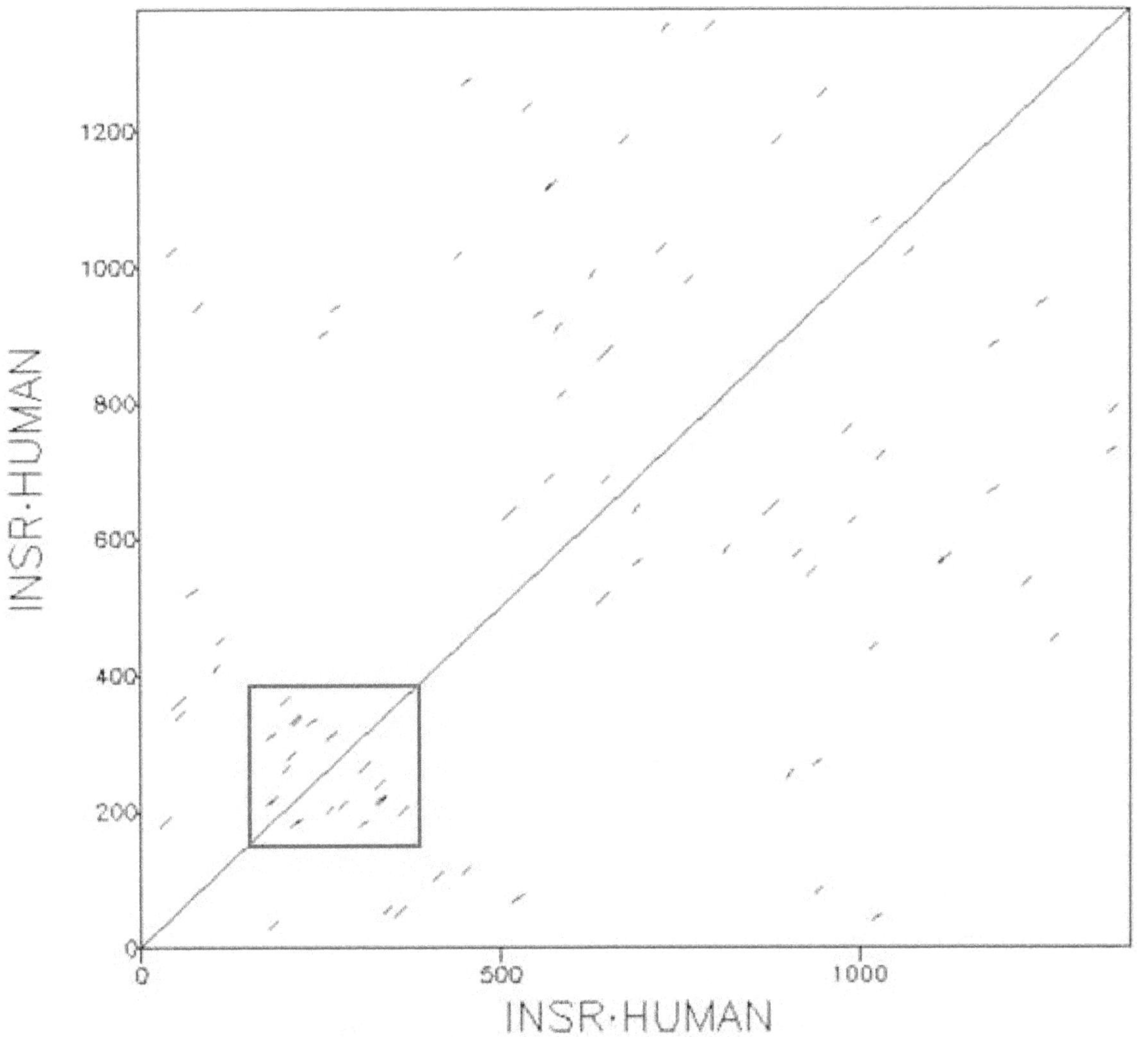

Figure 22. Dotmatcher plot of human insulin receptor protein versus itself. The rectangle indicates a region with higher density of matches, revealing composition bias in a cysteine rich region.

When looking into the properties of the INSR_HUMAN sequence using the InterPro[9] database (Blum et al., 2025) we find that it contains a domain called "Furin-like cysteine rich region" (**Figure 23**), covering positions 179-340. When analysing this sub-sequence using

[9] www.ebi.ac.uk/interpro

ProtParam, we found that cysteine, which is generally a rarely occurring amino acid, made up 16% of the sequence content in this region. It was thereby the most frequent amino acid in the region and approximately tenfold enriched compared to its general frequency in proteins (Nacar, 2023). Therefore, it again seems that a visible, but somewhat scattered, cluster of short matches corresponds to a highly biased amino acid composition.

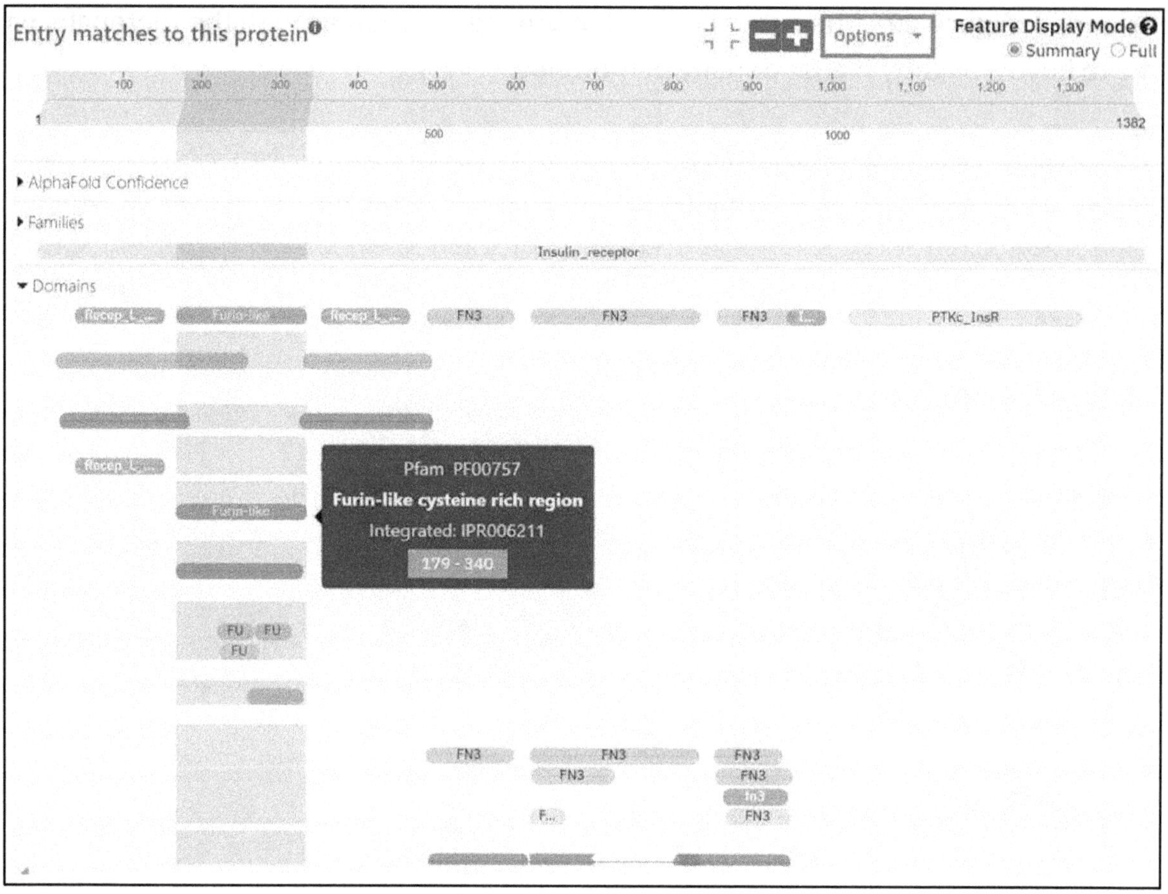

Figure 23. InterPro entry for the human insulin receptor protein, indicating the presence of various domains, including the Furin-like cysteine rich region.

As described in the discussion of the previous examples, regions with composition bias display a different visual pattern compared to repeat regions. However, as illustrated in **Figure 22**, it is unclear whether a cluster of matches due to composition bias will be distinguishable from random matches. This will naturally depend on the degree of bias – the more skewed the amino acid distribution is in comparison to the general background distribution, the more visually detectable should the cluster of matches be.

To further illustrate the effect of composition bias on dot plots, we used the tool RandSeq[10] to generate random protein sequences of the same length as INSR_HUMAN, i.e. 1382

[10] web.expasy.org/randseq

positions. We generated two random sequences, the first one with the standard amino acid composition and the second using the same biased composition as was found in the cysteine rich region covering positions 179-340 in INSR_HUMAN. The result, shown in **Figure 24**, shows a very distinct effect of the composition bias. As this experiment demonstrates, if the region with biased composition is sufficiently long and the bias sufficiently strong, the region should be detectable by visual inspection of a dot plot.

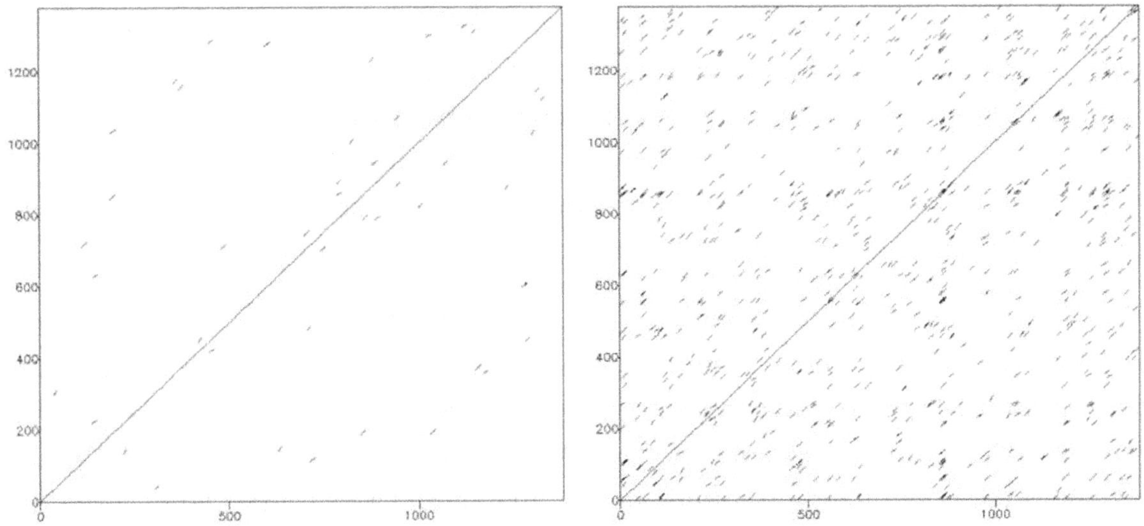

Figure 24. Dot plots of random protein sequences with standard amino acid composition (left) and composition bias corresponding to the cysteine rich region in INSR_HUMAN (right), showing that composition bias gives rise to a much larger number of sub-sequence matches.

Domain structure in protein sequence dot plots

In the previous chapter we discussed detection of repeats, which are usually referred to as a type of sequence feature. Sequence features are short, usually only a few amino acids long, and may confer a particular characteristic on the protein. Other examples of sequence features are active sites, binding sites, and post-translational modification sites, for example methylation sites. In this chapter we will instead turn our attention to domains, which are distinct structural or functional units in a protein that can fold, function and evolve independently (Wang et al., 2021). It is generally believed that one of the main drivers of evolution is the inheritance and reshuffling of domains (Doolittle, 1995; P. Bagowski et al., 2010; Schmidt & Davies, 2007) and a lot of effort has been made in bioinformatics research to develop methods to identify and characterise domains (Wang et al., 2021).

Many proteins contain multiple domains, and as an illustrative example we will here analyse the NCK1 protein, which contains three copies of the SH3 domain. The SH3 domain consists of approximately 50 amino acids and is involved in protein-protein interactions. Through this function, it is also involved in many biological processes, such as regulation of important cellular pathways, including cell proliferation, migration, and cytoskeletal modifications (Kurochkina & Guha, 2012).

The human NCK1 protein contains three copies of the SH3 domain, according to ScanProsite (**Figure 25**).

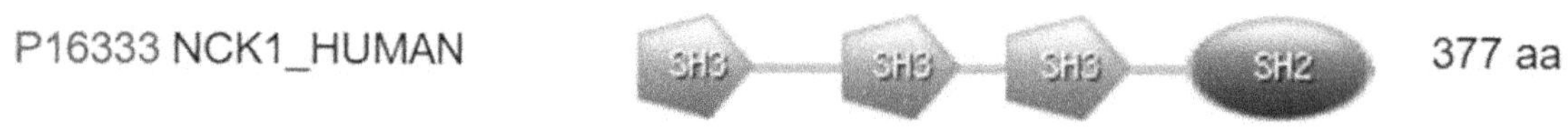

Figure 25. ScanProsite search results for NCK1_HUMAN, the amino acid sequence of the human NCK1 protein.

The positions of the three SH3 domains within the NCK1 sequence, according to the UniProt entry P16333, are 2-61, 106-165 and 190-252. These positions were indicated with bars on the x and y axes of the dot plot of the NCK1_HUMAN sequence against itself (**Figure 26**). The matches in the plot indicate that the two first copies of the SH3 domain are highly similar, except for the central part, where a short gap appears in the sub-sequence match. The third instance of the SH3 domain apparently differs substantially from the first and second copies in the first half of the region, since the match begins around position 220 instead of the expected starting position for the match, i.e. 190. We can also conclude that there is no detectable similarity between the SH2 and SH3 domains.

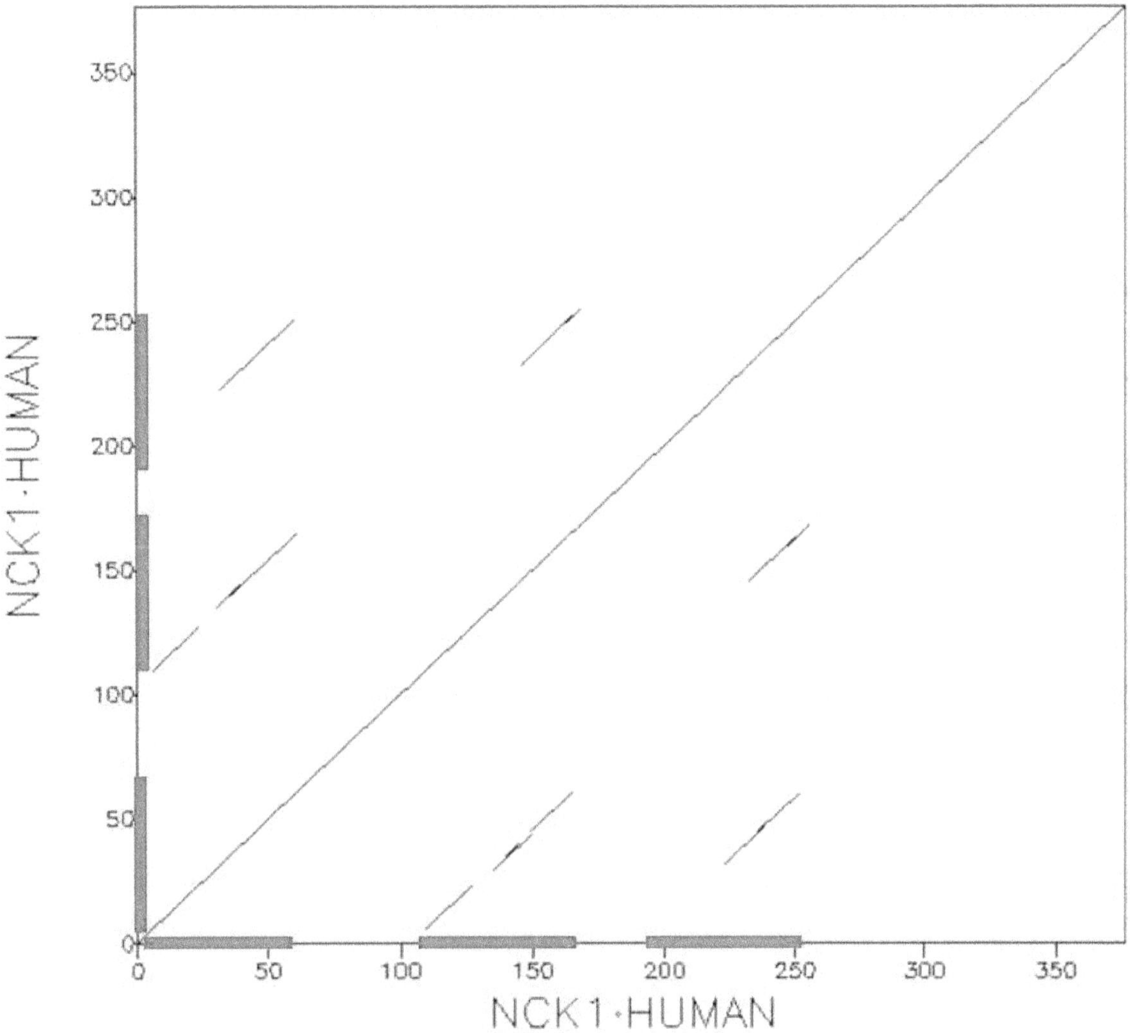

Figure 26. The sequence of protein NCK1_HUMAN plotted against itself, with bars on each axis indicating the positions of the three occurrences of the SH3 domain.

As concluding remarks regarding the examples shown in this example, we can make the following generalisations about how to interpret matches in a dot plot of a protein sequence against itself:

1. Repeat regions tend to appear as highly organised and regular parallel matches within a distinct region of the dot plot.
2. Composition bias also tends to appear as matches within a distinct region of the dot plot, corresponding to the extent of the composition bias region, but with more scattered and disorganised matches within this region.
3. Domain structure tends to appear as a few relatively long matches, corresponding to the number of copies of the domain.

Advanced algorithms for fast dot plot comparison of genome sequences

As stated in the introduction, the dot plot approach for comparing sequences has the advantage of being simple and fast, but it has the disadvantage of being imprecise. For pairwise sequence comparison at the gene or protein level, it is therefore in many cases preferred to use alignment tools instead, since these have the advantage of exactness (Needleman & Wunsch, 1970; Smith & Waterman, 1981). For a given pair of gene or protein sequences, and for a chosen scoring scheme, alignment algorithms based on the dynamic programming approach provide a guarantee of finding the optimal alignment (Altschul & Erickson, 1986). In this context, the “optimal” alignment is the one that gives the highest overall score out of all possible alignments for the given sequences and for the assumptions made, i.e. for the chosen scoring scheme. Provided that the scoring scheme has been chosen wisely, it is then further assumed that this alignment corresponds to how the two sequences relate biologically and that it correctly shows which positions have been preserved during evolution and which ones have been substituted. However, the exactness of alignment algorithms comes with the price of a much higher computational complexity, making them difficult to extend to the level of genome sequences. A further complication at the genome level is that genomes undergo rearrangements, such as translocations and inversions, affecting the genome sequence at a much larger scale than the base pair level mutations that we typically study when aligning gene or protein sequences.

Although there are tools for genome alignment (Armstrong et al., 2018), these are based on very complex algorithms and are highly costly to execute in terms of computer memory and processor runtime. It is therefore often desirable to use dot plots for an overview comparison of genomes at reasonable computational cost and to identify the large-scale rearrangements constituting the differences between the two compared genomes. However, even the comparatively fast approach of generating dot plots based on the sliding window approach requires some modifications to speed up the processing when dot plots are generated for genome comparison.

We will here discuss how the dot plot tool Gepard[11] can be used for dot plots at the gene or protein level and also scales up to the genome level, and the algorithm it uses for fast

[11] At the time of writing, the Gepard software was available for download from the web site for University of Vienna's Division of Computational Biology and the source code was available through GitHub.

cube.univie.ac.at/research/software-databases/genome-pair-rapid-dotter-gepard
github.com/univieCUBE/gepard

genome comparison (Krumsiek et al., 2007). In **Figure 27** we see a dot plot of the same pair of gene sequences as were compared with Dotmatcher in **Figure 11** and **Figure 12**. The most visually striking difference between the Dotmatcher and Gepard plots might be that Gepard places the origin in the northwest corner. When generating this plot, we uploaded the nucleotide sequence of the human S100 gene as the horizontal sequence and the corresponding rat gene as the vertical sequence. The dot plot was then created with default settings, which here means that the tool itself chooses to use either the sliding window approach and sets the window length itself, or it chooses the word-based approach and the word length. This is shown in the left panel in the screenshot under "Auto params", where "Window size" was automatically set to 25. The tool also automatically chooses a suitable substitution matrix based on the sequence input (in this case "DNA" as the choice for "Use matrix"). These parameters can also be set manually by the user, and the developers of Gepard state that the sliding window approach should only be used for sequences shorter than approximately 10,000 characters.

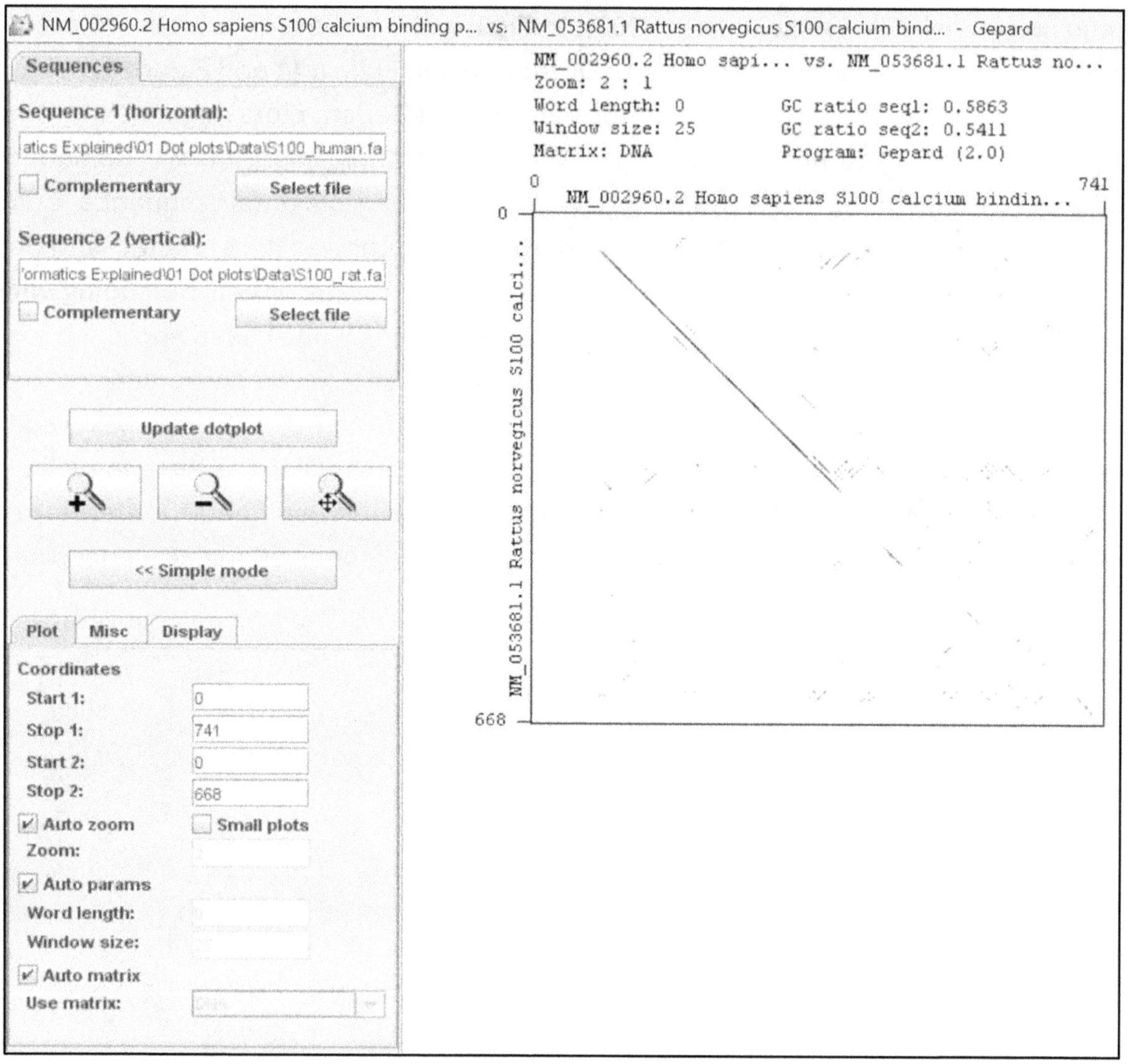

Figure 27. Dot plot of mRNA sequences from the human (vertical axis) and rat (horizontal axis) S100A3 gene, encoding the S100 calcium-binding A3 protein. Generated with Gepard, using automated parameter settings. In this case the tool, based on the input sequence type and length, chose to apply the sliding window approach with window length 25 and the "DNA" substitution matrix.

When it is run with the sliding window approach, Gepard will produce essentially the same results as we saw earlier for Dotmatcher, and for such short sequences as in this example (741 vs 668 nucleotides), the calculation of the plot will take only a fraction of a second. For longer sequences, however, such as whole chromosomes or genomes, the sliding window approach is too slow. In **Figure 28** we see an example where Gepard was used to compare the genome sequences of two different strains of *Mycoplasma genitalium*. This organism has one of the smallest known bacterial genomes, consisting of approximately 0.5 million bp. Already for such a small genome, we do however run into the problem of long processing times. If we use the sliding window technique, scores will need to be calculated for approximately $500{,}000^2$ = 250 billion matrix cells, where each window score is the sum over the window length. Obviously, this is a prohibitively large number of calculations and the Gepard software will therefore automatically switch to using the more efficient word-based approach. If the user tries to manually set parameters that would lead to an excessively long computation time, the software will show a warning and suggest considering a more time-efficient setting.

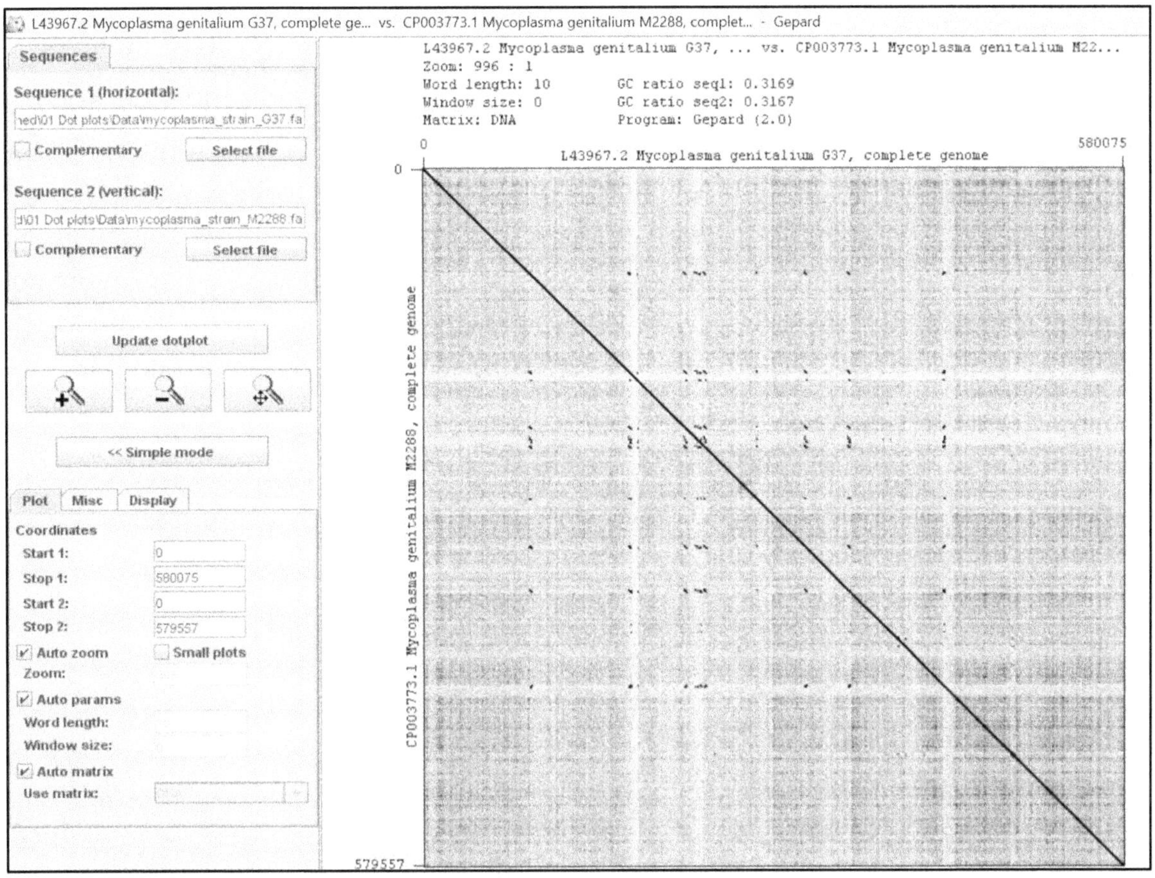

Figure 28. Gepard dot plot comparison of the full genome sequences of two *Mycoplasma genitalium* strains. Genome sequences were downloaded from Genbank entries with accession numbers L43967 and CP003773. Gepard was run with automated choice of parameter settings.

When setting a word length instead of a window size, Gepard uses a word index to quickly locate matching sub-sequences. It does this by first finding all unique "words", i.e. sub-sequences, of a given length in one of the sequences, building an index to keep track of the locations of these words, and then it uses this index to quickly find where each word of the other sequence matches with the former. For this approach to work efficiently, it is necessary to use a suitable data structure for storing the indexed words, a method of creating this data structure, and a method for utilising the data structure to find the matches. We will in the following first describe the data structure, called a suffix array, then discuss how it is created and how it is used when finding the matching sub-sequences for the dot plot.

Informally defined, a suffix array is a list of starting positions of suffixes of a text string, sorted in such a way that the suffixes appear in alphabetical order. **Figure 29** gives a brief schematic overview of how a suffix array is generated and how it is used. We here create a suffix array for a string consisting of 15 nucleotides. A suffix is any sub-sequence that ends with the final `t` at position 14 in the string. The suffixes are then sorted in alphabetical order, starting with `atatt` and ending with `tt`. In the suffix array we also note the starting position in the original string for each suffix, for example position 10 for `atatt` and position 13 for `tt`. The advantage of this data structure is that any substring from the second sequence, such as the substring `atg` in the beginning of the sequence `atgcgcgtagctagc` in the example, can rapidly be matched to the identical substrings in the suffix array by binary search. A binary search means that we repeatedly divide the search space in halves. This search algorithm assumes that the elements in the search space, in this case the suffixes, have been sorted. Provided that this is the case, we can for each step check the element at the midpoint to decide if the element we are looking for is located before or after the midpoint. This is illustrated by the steps 1, 2 and 3 in **Figure 29**.

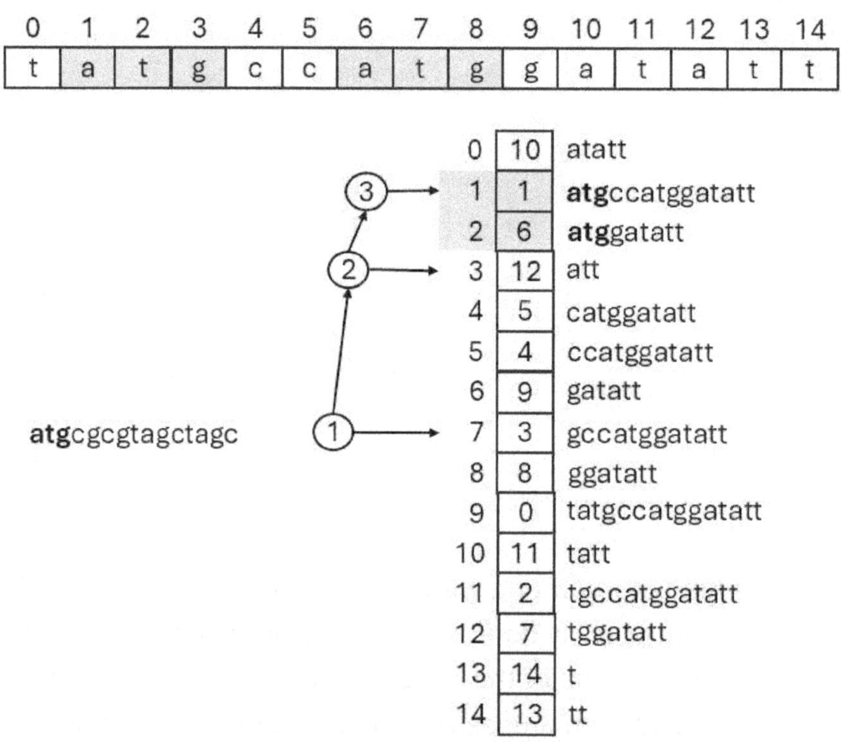

Figure 29. Toy example overview of how suffix arrays are generated and used.

In the example in **Figure 29**, we are looking for suffixes beginning with `atg`. In the first step we compare with suffix 7 in the middle of the array and note that `atg` should alphabetically occur before any suffix beginning with `g`. This reduces the size of the search space to half, from 16 to 8 elements. In the second step we compare `atg` to suffix 3, which is in the middle of the upper half of the array and note that `atg` appears before `att` in alphabetical order. In the third step, we already find the part of the array where all suffixes beginning with `atg` are located. We can now directly see that the `atg` substring from the query sequence matches with substrings beginning at positions 1 and 6 in the indexed sequence. In general, the number of steps used by a binary search algorithm for a sorted search space with *n* elements is approximately *log2 n*. For the toy example shown here, with a sequence of length 16, this is 4 steps. This can easily be derived from the fact that we cut the search space successively in halves, meaning that an initial search space of size 16 is in four steps cut down to 8, 4, 2 and finally 1 element, which is the one being searched for.

Going back to the application example of comparing two *Mycoplasma genitalium* genome sequences to each other, we can note that with a naive approach of finding matching substrings, the computational complexity is very high. We could take each substring, i.e. word, from one sequence and search linearly through the complete length of the other sequence to find all locations where it matches. With genome sequences of approximate length 0.5 million bp, this would be an extremely time-consuming approach since we would need to evaluate half a million potential match positions for each word. If instead using a suffix array representation for one of the sequences, then matching each word of the other sequence to the suffix array, we would go through only log2(500,000) ≈ 19 steps of binary search on average to find all matches for each word. This very significant reduction in computational complexity, due to the usage of suffix arrays, makes the word-based approach a suitable alternative to the window-based approach when generating dot plots for genome-size sequences. The extra time needed to generate the suffix array is, with a wide margin, made up for by the very large reduction of search time when matching words to the sorted suffixes.

It should be mentioned that very efficient algorithms have also been developed for the task of generating suffix arrays, which it is beyond the scope of this text to discuss. For a detailed description of such methods, see, for example, Shrestha et al. (2014).

When the mycoplasma example (**Figure 28**) was generated using automated parameter choices, the Gepard software set the word length to 10, meaning that it will search for every sub-sequence of length 10 from one sequence within the suffix array generated from the other sequence. Obviously, if the word length is increased to a higher value, it will search for fewer words, which will reduce the search time. Increasing the word length can therefore be considered as a means of reducing the computational time when comparing very long sequences. However, increased word length will, on the other hand, reduce the sensitivity since shorter matches will not be detected. In **Figure 30** we can see the results when setting word length to increasingly high values. Increasing the word length from 10 (left) to 100 (middle) has the effect of filtering out short and mostly irrelevant matches, which showed up as a generally grey background "noise" in the first plot. On the other hand, it can be argued that some of these shorter matches show regions where there are repeats, which may be relevant to identify. When setting the word length to 100, both the noise and the short repeats were absent from the plot, but it preserved the main diagonal. This shows that the two genomes are overall highly similar, without any detectable major rearrangements. When increasing word length further to 1000 (right), the main diagonal starts breaking up into pieces. Here, the gap regions can be interpreted as indicating regions where the two genomes have lower overall similarity than in the regions covered by matches.

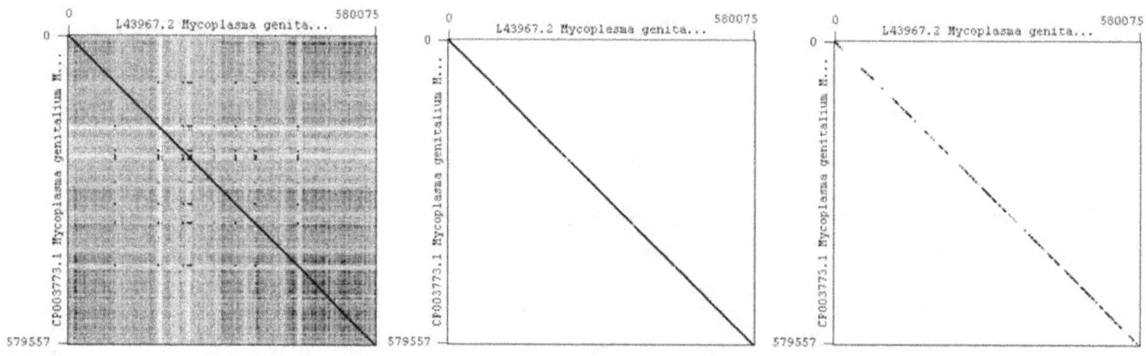

Figure 30. Gepard dot plots of *Mycoplasma genitalium* genome sequences with word length set to 10 (left), 100 (middle) and 1000 (right).

We now turn to an example where slightly larger genomes are compared. The bacterium *Haemophilus influenzae* is responsible for a wide range of infectious and inflammatory conditions in humans, including pneumonia, meningitis, and bloodstream infections. Its genome consists of a singular circular chromosome, approximately 1.8-1.9 million bp in length, containing around 1600 protein-coding genes. Since the first *Haemophilus* genome was fully sequenced and published in the mid-1990s (Fleischmann et al., 1995), the genomes of many different *H. influenzae* strains have been completed, making it possible to study how new strains evolve by genomic rearrangements. The dot plot in **Figure 31** shows an example of such a comparison, where the genome sequences of the strains PittGG and PittEE (Hogg et al., 2007) are plotted against each other.

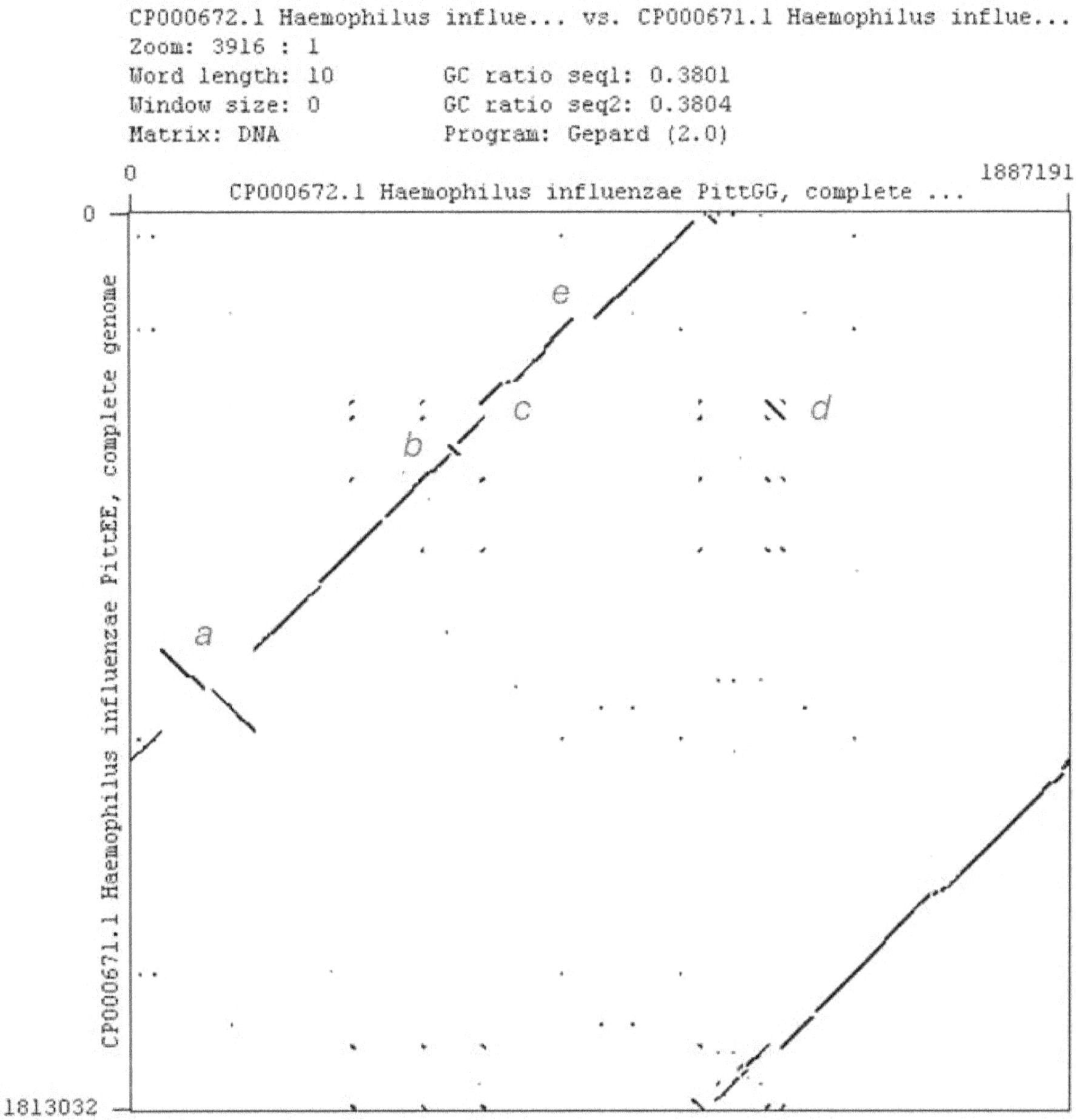

Figure 31. Gepard dot plot comparison of the genomes of *Haemophilus influenzae* strains PittEE (Genbank accession number CP000671) and PittGG (CP000671). Letters indicate types of genome rearrangements, such as inversion (*a*, *b*), inversion plus translocation (*c*, *d*) and insertion/deletion (*e*).

The first observation to make about the dot plot in **Figure 31** is that the main diagonal does not extend from one corner of the matrix to the opposite corner. Instead, it appears to consist of two parts – a longer one in the northwest region of the plot and a slightly shorter part in the southeast region. This can be explained by the fact that the chromosome *in vivo* is circular, whereas the genome sequence, as it appears in the Genbank database in text format, has a beginning and an end. For the two database entries, the start- and endpoints of the sequence do not coincide, which results in the wraparound effect in the dot plot.

For a circular genome, the "beginning" can be defined as the location of the origin of chromosomal replication (Ekundayo & Bleichert, 2019), but genome entries in Genbank are not synchronised in terms of the location of the origin of replication within the text string representing the sequence. When using the Ori-Finder[12] tool (Dong et al., 2022), we found that it predicted the origin of replication region to begin at 11,048 nt in the PittEE sequence and at 1,161,606 nt in PittGG. This, rather large, difference of 1,150,558 nucleotides corresponds well with the amount of "shift" in the plot, as indicated by the vertical line in **Figure 32**.

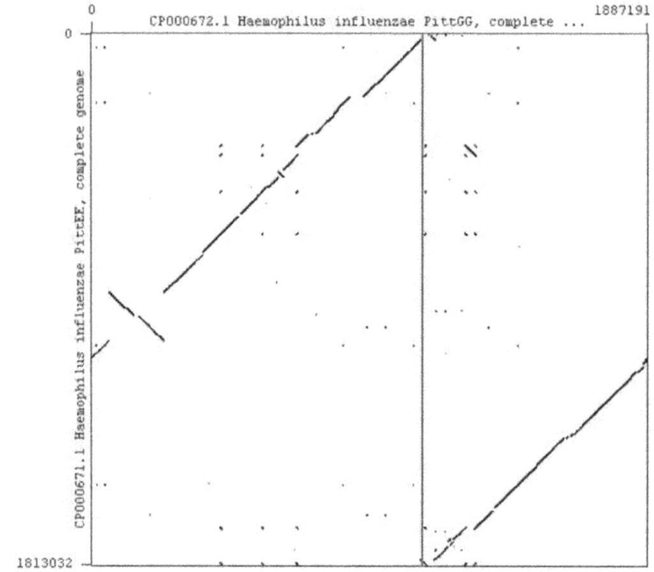

Figure 32. The same Gepard dot plot as in **Figure 31** with a vertical line added, indicating the point along the *x* axis that corresponds to the amount of difference between the predicted origins of replication for PittEE and PittGG (1,150,558 nt).

A second observation is that the two main diagonals are orientated from the upper right to the lower left, which means that one of the sequences is matched in the reverse orientation. This is of no consequence for the analysis and simply reflects the orientation of each sequence in the database entry. For Genbank sequences, when observed in a web browser, one can select the setting "Show reverse complement" as display option in the "Customize view" setting before downloading the FASTA entry (see **Figure 33**). Doing that for one of the sequences in the dot plot in **Figure 31**, we would see the main diagonal displayed in the opposite direction, i.e. from upper left to lower right. The same effect can also be achieved by clicking the "Complementary" check box in Gepard when uploading one of the sequences.

[12] tubic.org/Ori-Finder2022

Regarding the actual comparison of genomes, the letters in **Figure 31** indicate areas where the plot identifies possible genome rearrangements. Letters *a* and *b* indicate inversions, where a segment of the chromosome has been inverted in its original location. Letters *c* and *d* together indicate a case where a segment seems to have been both inverted and translocated, i.e. excised from the chromosome, inverted, and inserted at a different location. Finally, *e* indicates a possible deletion or insertion, where a part of the PittGG sequence does not have any correspondence in the PittEE sequence. It cannot be determined only from observing the dot plot whether this difference appeared due to insertion in one chromosome or by deletion in the other. It can be noted that the PittEE genome sequence is approximately 70 kbp shorter than the PittGG sequence (1,813,032 versus 1,887,191 bp). The observed insert/delete could be the reason for this length difference, although other explanations are, of course, possible.

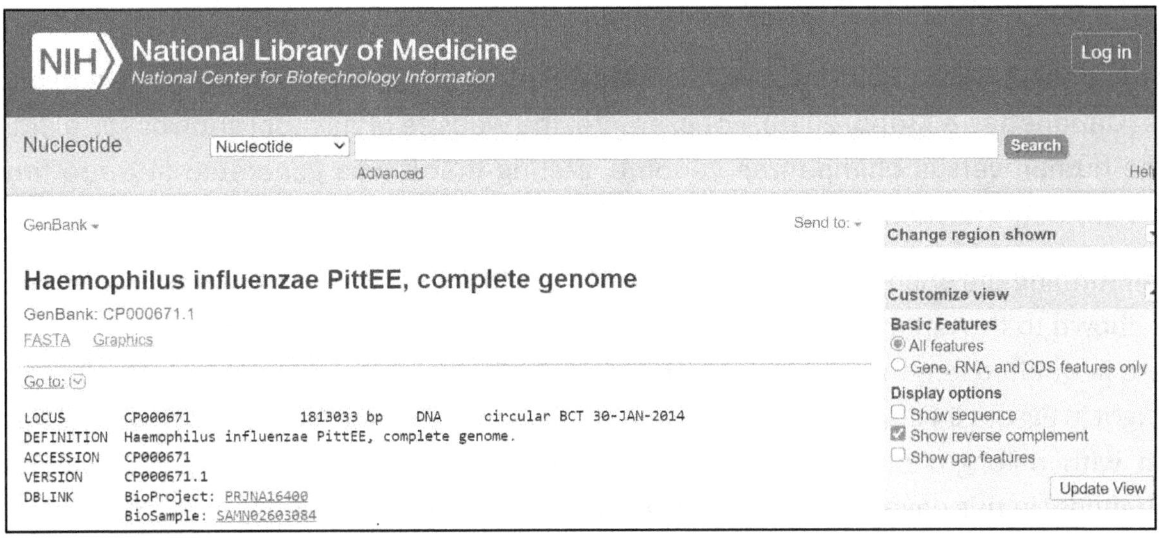

Figure 33. Setting the display options in Genbank to show the reverse complement of a sequence. In the "Customize view" menu we have here selected "Show reverse complement" in the "Display options".

Ultra-fast algorithms for analysis of large genomes

Using the word-based approach is computationally efficient and can work well, as we saw in the previous section, for comparison of genomes of lengths at the level of millions of base pairs. However, for larger genomes even this approach becomes too time consuming. Considering the human genome as an example, with its approximately 3 billion base pairs, mapping each word of this sequence to a suffix array would entail execution of up to 3 billion such mappings. In addition, the initial step of building the suffix array would be very time consuming and would take up large amounts of memory just for storing the array. For these reasons, even more efficient algorithms and data structures are needed to generate dot plots for comparison of eukaryotic genomes.

The program D-GENIES can be used to create dot plots for genomes of any size in reasonable time (Cabanettes & Klopp, 2018). For example, the website of this tool showcases a dot plot of the human versus chimpanzee genome, stating that it was generated in a run time of approximately one hour using up to 34 GB of RAM memory.

When running the online version of D-GENIES[13] on the *H. influenzae* example we get the dot plot shown to the left in **Figure 34**. The runtime was reported as 1 second for this example, where sequence lengths were approximately 1.8 Mbp. For a larger example, the dot plot to the right in **Figure 34** compares chromosome 1 of *Homo sapiens* with that of *Pan troglodytes*, each with a length of over 200 million base pairs, i.e. approximately 100-fold longer sequences. In this case, runtime was slightly less than one minute.

[13] dgenies.toulouse.inra.fr

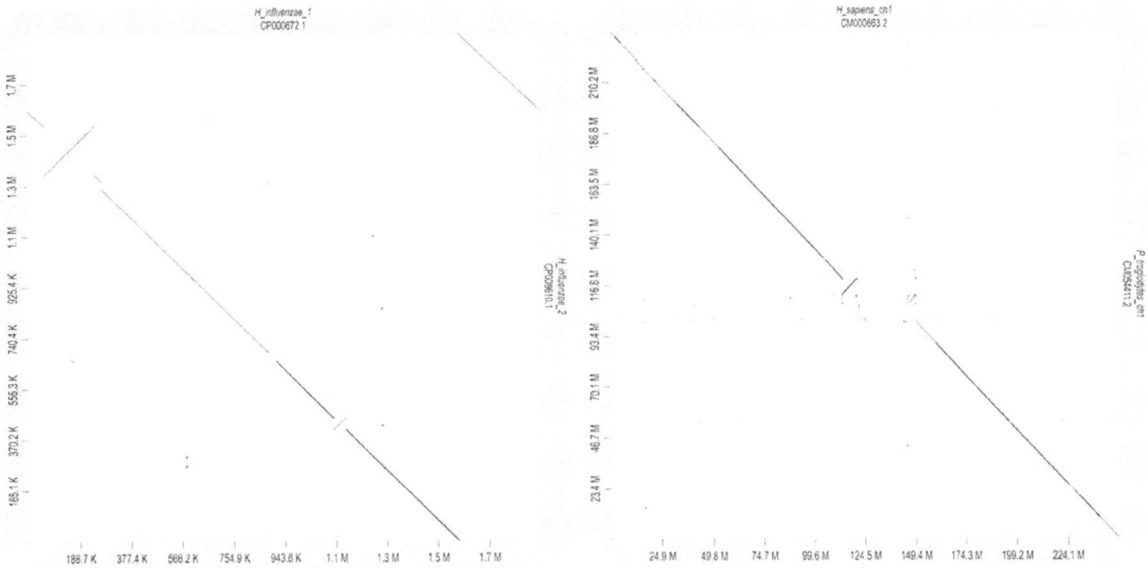

Figure 34. D-GENIES dot plot comparisons of genomes. Left: *Haemophilus influenzae* strains PittEE and PittGG (Genbank accession numbers CP000671 and CP000671). Right: Chromosome 1 of *Homo sapiens* and *Pan troglodytes* (CM000663 and CM054411).

To achieve its superior processing speed, D-GENIES applies an approach referred to as sketching (Das & Schatz, 2022; Ndiaye et al., 2024; Zheng et al., 2023). In general, sketching is used to create a reduced representation of a large dataset in such a way that the important properties of the original data set are preserved. If such a reduction can be achieved, then the reduced data set can be used instead of the original one to achieve the same result with less processing. For the algorithm used by D-GENIES, the reduction of the data set is done by first considering all *k*-mers of the sequence, i.e. all sub-sequences of length *k*, where each *k*-mer overlaps with the next by *k*-1 positions. From the full set of *k*-mers, the algorithm chooses a considerably smaller subset which, if strategically chosen, will be sufficient for identifying the matching regions between two genomes in a dot plot. The smaller subsets of *k*-mers are in this context called minimizers. Minimizers were first proposed as a method of data compression for large-scale genome assembly projects (Roberts et al., 2004) and have subsequently been used in algorithms for mapping reads to genomes and in *de novo* assembly methods (Li, 2016, 2018).

Figure 35 illustrates schematically the concepts of *k*-mers and minimizers. The original representation of the sequence, here called the sequence space, is a string *S*, consisting of letters representing the nucleotides. The size of the sequence space can be written as |*S*|, representing the number of characters, i.e. the length of the sequence. The *k*-mer space consists of all substrings of length *k*, where each pair of contiguous *k*-mers overlap by *k*-1 positions. Naively storing all *k*-mers would take up more memory than the sequence itself, since it would require a space of approximately |*S*|**k*. However, we should note that there will be many copies of the same k-mer in the sequence, but we will only need to store each k-mer once, together with an index pointing to its locations in the sequence. For the toy example in the figure, with k=3, there is a total maximum of 4^3 = 64 (in general 4^k for nucleotide sequences) unique k-mers. Among the 18 k-mers in this example, there are two, AGC and TGA, that occur twice. In memory we therefore need to store only 16 unique k-mers with indices showing their locations in sequence space.

Sequence space	TGACAAGC...AGCACCCT...GCTCTTGA...			
k-mer space	TGA GAC ACA CAA AAG AGC	AGC GCA CAC ACC CCC CCT	ACT CTC TCT CTT TTG TGA	{ACA,ACC,AGC,CAA, CAC,CCC,CCT,CGT, CTC,CTT,GAC,GCA, GCT,TCT,TGA,TTG}
Minimizer space	ACA	ACC	CTC	{ACA,ACC,CTC}

Figure 35. Overview of the concept of minimizers, with example loosely based on Figure 1 in the review article by Ndiaye et al. (2024).

It can here be noted that the number of possible k-mers increases exponentially with k, so that it will exceed $|S|$, but for high values of k many of the possible k-mers will not occur in the sequence. Obviously, only the actually occurring k-mers are necessary to store in memory, together with the index showing their locations. For many applications in bioinformatics that use k-mer representations, the choice of k becomes an issue of achieving a balance. For a dot plot tool, a low value of k would result in high sensitivity for short matches, combined with a long processing time. A higher k would speed up the processing, since fewer potential matches would be evaluated, but the gain in speed would come with the risk of missing some relevant matches, i.e. lower sensitivity.

For bioinformatics applications handling very large datasets, the size of the k-mer space becomes impractical and, in many cases, it provides more detail than what is necessary. In such cases minimizers can be applied to reduce the memory demand. The general idea of minimizers is to use only a subset of the k-mers, chosen in such a way that memory space is significantly reduced, but also so that the original information is retained, thus ensuring that the same sub-sequence matches will be identified for the dot plot as when using the full k-mer space. In **Figure 35** the representative k-mers in the minimizer space, i.e. the "minimizers", were arbitrarily chosen. In practice, algorithms based on minimizers choose the minimizers strategically to achieve both exactness and minimal storage space. As a practical example of this, Roberts et al. (2004) reported that the 200 Gb k-mer database for the *Rattus norvegicus* reference genome could be reduced to a 20 Gb minimizer database without loss of information. Thus, a genome assembly problem that previously required a computer cluster was made feasible to be run on a desktop computer and even within reach of being processed on a powerful laptop computer.

Figure 36 illustrates a method for choosing minimizers strategically. The left part of the figure shows that each row below the sequence *S* represents a "window", where each window corresponds to *w*+*k*-1 consecutive *k*-mers. In the example we are using *w* = *k* = 3, resulting in a window length of 5. Considering the window in row 1, it consists of the three overlapping *k*-mers TGA, GAC and ACA, which together represent the five initial nucleotides in the sequence, i.e. TGACA. In each window we now choose the "smallest" *k*-mer as the minimizer of the window. What is meant by "smallest", can be adjusted depending on the type of sequence data and the application (see Roberts et al. (2004), section 2.4, for a discussion of the merits of different orderings). For this example, we will define "smallest" as the *k*-mer sequence that comes first in alphabetical order. Based on this reasoning, ACA will be the minimizer for the first and second windows. For the third window, AAG is the "smallest" *k*-mer, and it will be chosen as minimizer also for the fourth and fifth windows. Continuing this procedure, we will generate a total of five minimizers for the sequence *S*, as illustrated in the right-hand part of the figure.

S	TGACAAGCAGCACCCT	*S*	TG**ACAAGCAGCACC**CT
1	TG**ACA**		**ACA**
2	G**ACA**A		**AAG**
3	AC**AAG**		**AGC**
4	C**AAG**C		**AGC**
5	**AAG**CA		**ACC**
6	**AGC**AG		
7	GC**AGC**		
8	C**AGC**A		
9	**AGC**AC		
10	GC**ACC**		
11	C**ACC**C		
12	**ACC**CT		

Figure 36. Minimizers for the string *S* when *w* = *k* = 3, corresponding to a window size of *w*+*k*-1 = 5 and using lexicographic ordering to choose the "smallest" *k*-mer for each window. As indicated with bold face in *S* to the right, all nucleotides are covered by the five minimizers, apart from *w*-1 = 2 nucleotides in the beginning and end of the string.

Note that AGC is chosen as minimizer for two different positions in *S*. In general, for a long sequence, we will of course use the same minimizer for many different windows, and this is one of the reasons why minimizer space is much smaller than the full *k*-mer space. For the toy example in the figure, with sequence length |*S*| = 16, the reduction is from |*S*|-*k*+1 = 14 *k*-

mers to four unique minimizers. Roberts et al. (2004) reported that using $w = k = 20$ in real applications, such as genome assembly, the reduction is approximately 10-fold.

An important aspect of the method outlined above is that setting $w \leq k$ guarantees that no gaps occur between the minimizers, thus ensuring that no information is lost. More specifically, there is no loss of information needed for mapping sub-sequences to the original sequence, *S*. At the ends of the sequence, at most *w*-1 nucleotides may be uncovered. In our toy example in **Figure 36** two nucleotides were uncovered in the beginning and end of *S*. This minor problem can easily be solved by adding extra end-minimizers.

D-GENIES uses the algorithm minimap2 (Li, 2018) for generating the set of minimizers from one sequence and for mapping parts of the other sequence to the first by matching with minimizers. This algorithm first creates a set of minimizers, according to principles described above, which are then stored in a hash table. For each element in this table, the key is the hash of a minimizer, and the value is a list of that minimizer's locations in the sequence. When mapping a part of the other sequence to the first sequence, it uses a procedure described as seed-chain-align (Li, 2018). In the seed phase, it finds the locations of matching minimizers, which represent short exact matches that can be used as "seeds". These are also referred to as "anchors". It then identifies sets of collinear seeds/anchors that fulfil certain criteria, making them promising candidates for longer matches. A set of collinear anchors is referred to as a "chain". Finally, in the "align" phase, it attempts to fill in the gaps between the anchors of a chain and also to extend the match beyond the beginning of the first anchor and beyond the end of the last anchor in the chain. During the aligning, mismatches and gaps are allowed, but the algorithm will find the alignment that gives the best score using a function that sums up the total number of nucleotide identities and applies a certain penalty for each gap introduced (Li, 2018).

To illustrate the seed-chain-align procedure in a bit more detail, let us first consider how a hash table of minimizers is constructed. Briefly described, a hash function is a function that takes an input value (in our case a minimizer) and maps it to a value that can be used as an index for elements in a table. There are many methods for creating hash values, but for the sake of brevity we will here just consider one method that works for our example data. **Figure 37** shows how a hash index can be generated for the set of minimizers we derived in **Figure 36**. We first decide on a coding for the nucleotides, in this case A = 0, C = 1, G = 2, T = 3. Secondly, we choose a hash function for converting *k*-mers to index values. We have here chosen a so-called polynomial rolling hash function. The general definition of such a function, for a string *s* of length *n*, is

$$\text{hash}(s) = \sum_{i=0}^{n-1} s[i] \cdot p^i \bmod m =$$

$$= s[0] + s[1] \cdot p + s[2] \cdot p^2 + \ldots + s[n-1] \cdot p^{n-1} \bmod m$$

where the values of p and m are positive integers and chosen depending on the data. For the example in **Figure 37** we have chosen p = 4 and m = 13 (to make our example similar to the one in Figure 4 of Ndiaye et al. (2024)). A larger value of m increases the size of the hash table and leads, therefore, to fewer cases of different k-mers mapping to the same value, but it will, on the other hand, make the algorithm slower. In our example, with k = 3 and a four-letter alphabet, there are 4^3 = 64 possible k-mers and 13 elements in the hash table, which will obviously lead to a number of collisions. For example, CAC and ACA will both map to index 4.

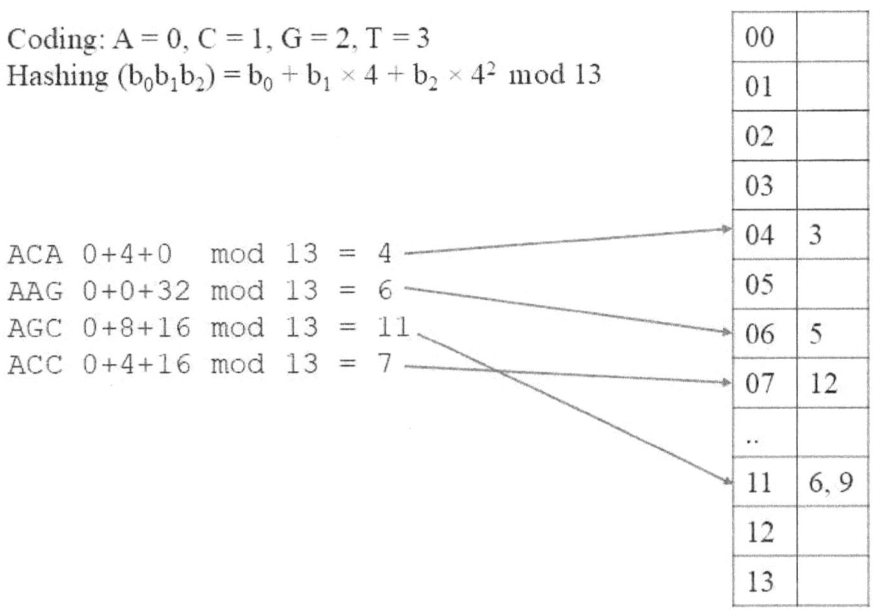

Figure 37. Generating a hash table for minimizers and their positions in the sequence S, using the rolling polynomial method. For each minimizer k-mer the hash function is applied to calculate its index in the table, then the corresponding table entry stores a list of the locations of that k-mer in the sequence. For example, ACC generates the hash 11, and this element in the table will then store the locations 6 and 9, representing the locations of the two occurrences of ACC in the sequence S.

There are various methods for what to do in cases of "collisions", i.e. different k-mers mapping to the same hash value. A straightforward method is to use a second hash function with different values of p and m to resolve such cases. The "collision" elements in the first hash table will then contain links to the corresponding elements in the secondary table. More advanced methods aim to achieve a so-called minimal perfect hash, i.e. a hash function that produces a table that is minimal for the data set (has no empty elements) and yet is collision-free (Limasset et al., 2017).

Once we have the hash table, it is an easy task to find the "seeds" or "anchors" when matching a part of a sequence to the table. For example, when matching the sequence S_2 = C**ACA**GGAGG**ACC**TT to the sequence S from **Figure 36**, for which we generated a set of minimizers and indexed them in the hash table shown in **Figure 37**, we would get the anchors ACA and ACC, since these *k*-mers from S_2 match two of the minimizers. The hash table shows that the positions of these anchors in *S* are 3 and 12. In this way, the algorithm will first identify all anchors, then sort them according to their positions in *S*, identify chains of colinear anchors located within a certain maximum distance of each other, and finally – in the "align" phase – attempt to join them into a longer match using an alignment procedure.

For our toy example, the alignment phase entails comparing the sequences *S* and S_2 in the region between the two anchors and applying a dynamic programming technique to find the best scoring alignment for this region (Eddy, 2004). When doing so, it allows mismatches and the introduction of gaps. The resulting alignment might look as follows, with the anchors marked in bold font:

C**ACA**GG-AGG**ACC**TT
G**ACA**AGCAGC**ACC**CT

In this case, the aligned region between the two anchors contains three identities, two mismatches and one gap. Chains of anchors that can be joined with sufficiently high-scoring alignments, considering the percentage of identical positions and the total penalties for gaps, are reported as matches which will be included in the dot plot.

Application example 1: Comparing multiple clustered genomes

In this series of application examples, we will look at different ways that dot plots have been used in research. Each example will also be recreated, fully or partly, to show step by step how dot plot tools are used in practice. The reader is encouraged to follow the steps for each example as a way of practising and gaining experience in using dot plot software. For each example we will briefly describe the research background and the purpose of using dot plots as part of the study, before we show how parts of the results presented by the authors can be recreated. It should be emphasised that these are just a few examples out of a vast literature. The article that presented the Gepard tool (Krumsiek et al., 2007) has to date[14] been cited in 640 scientific publications, according to the Dimensions citation database (Singh et al., 2020), and the D-GENIES article (Cabanettes & Klopp, 2018) has been cited 581 times. There are, in other words, many additional application examples out there in the scientific literature that the reader is encouraged to explore in addition to the ones presented here.

In a study published in 2025, Kok and co-workers performed a comprehensive analysis of the genomes of 96 *Paenibacillus larvae* bacteriophages (Kok et al., 2025). This phage is of great practical interest since the *P. larvae* bacterium infects the domestic honeybee (*Apis mellifera*), causing the highly infectious American foulbrood disease. Outbreaks of this disease cause devastating consequences for farmers since infected bee hives must be burnt to prevent further spread. A lot of research has therefore been performed on possible treatments, with phage therapy being one of the candidate approaches (Brady et al., 2017). If successful, phage therapy may be preferable since antibiotics have several disadvantages, such as the risk of development of antibiotic resistance, disruption of the gut bacteria balance in the bees leading to increased susceptibility to other infections, and the possibility that antibiotic residues in honey may pose health risks for consumers (Shoaei et al., 2024).

In their study, Kok and co-workers presented 26 newly sequenced, assembled and annotated *P. larvae* phage genomes (Kok et al., 2025). Noting that their contribution brought the number of known *P. larvae* phage genomes to a total of 96, they proceeded to perform a comparative analysis of all these genomes. Characterising phage genomes is important for the potential use of phages to treat infectious diseases. In particular, it is crucial to identify genes that enable the phages to perform the function desired for the therapy, i.e. genes making them lysogenic, but also genes that could make them unsuitable for use in therapy,

[14] March 2025

for example antibiotic resistance genes. In addition, it could be necessary to adapt the choice of phage to the strains of *P. larvae* that infect the bee populations in different regions.

In short, the methods used in the Kok et al. study were to first make pairwise alignments of all pairs of genomes and compute their percent aligned nucleotide identity (PNI), then generate a phylogenetic tree showing how all 96 genomes relate to each other, followed by genome comparisons using dot plots to identify both conserved and divergent regions between related genomes. Since bacteriophages are viruses, their genomes are extraordinarily short – in this case approximately 40 thousand nucleotide base pairs, containing only about 80 genes. Regarding the 26 newly sequenced phages presented in the article, two distinct clusters were identified. The genomes of 23 phages were all very similar to each other, with PNI ranging from 97.3% to 99.9%. The remaining three genomes formed a separate cluster which was clearly distinct from the main cluster, but also had lower PNI values when compared to each other. The dot plot comparison of the 26 newly sequenced genomes is shown in Figure 3 in the article (Kok et al., 2025). In **Figure 38** we have reproduced parts of the plot, including nine of the 26 genomes. We will here first describe how the reproduced plot was generated and after that discuss its interpretation and the biological insights gained from it.

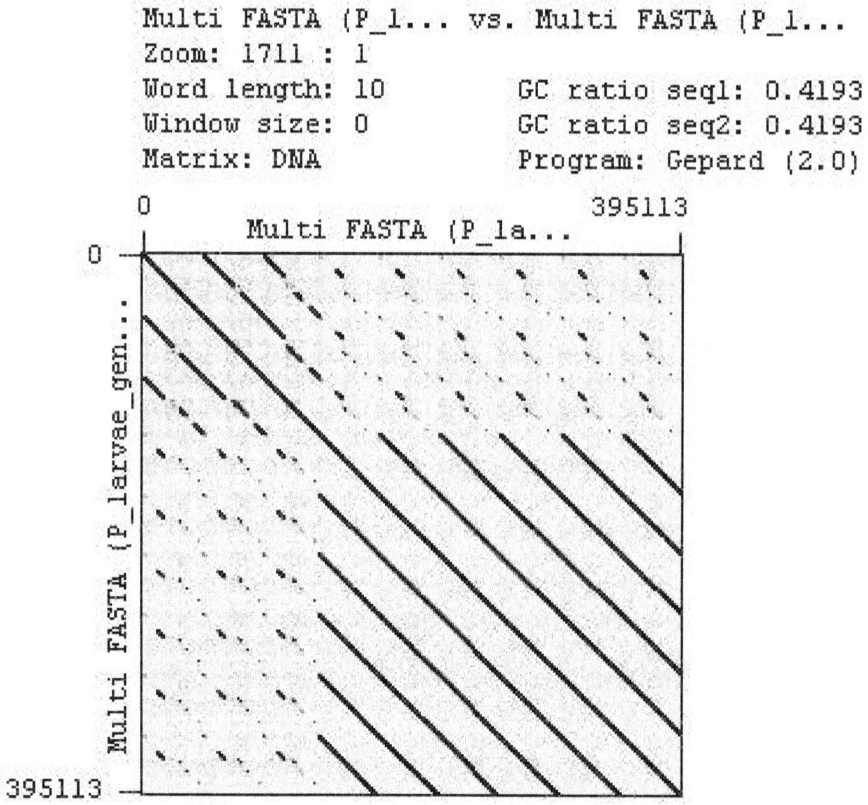

Figure 38. Gepard dot plot comparing nine *P. larvae* genome sequences.

To find the genome sequences used in the Kok et al. (2025) study, we first went to the online version of the article[15] and searched for the Supplementary Materials section. We there found a link to a compressed archive containing, among other files, a table that shows the Genbank accession numbers of all genome sequences used in the study. From this table (Supplementary Table S1) we selected the nine entries shown in **Table 2**.

We then created a text file containing only the accession numbers, with each number on a separate line. To find the nine entries in Genbank, we uploaded this text file through the Batch Entrez download interface[16] and retrieved the nine records (**Figure 39 a**). On the Genbank search results page, we used the "Send to" link to generate a multi-FASTA text file containing the nine genome sequences listed in order, with each sequence preceded by a header line (**Figure 39 b**). This multi-FASTA file was then uploaded in the Gepard dot plotter tool both as sequence 1 (horizontal) and sequence 2 (vertical) and plotted against itself. The resulting plot in **Figure 38** shows the same overall patterns of similarity as Figure 3 in the article (Kok et al., 2025), except that only nine of the 26 sequences are included.

Table 2. Sequences used for the dot plot shown in **Figure 38** with GenBank accession numbers for reference to each sequence entry.

Phage Name	Cluster	Genus	GenBank Accession No.
Callan	Harrison	Fernvirus	OP503989
Dash	Harrison	Fernvirus	OP503990
Lilo	Harrison	Harrisonvirus	OP503991
Carlos	Vegas	unclassified*	OP503973
Rosalind	Vegas	unclassified*	OP503984
GIW2016	Vegas	unclassified*	OP503977
ABAtENZ	Vegas	unclassified*	OP503968
AJG77	Vegas	unclassified*	OP503969
ApiWellbeing	Vegas	unclassified*	OP503970

Along the horizontal and vertical axes of the plot in **Figure 38**, genomes appear in the same order as they are listed in **Table 2**. The most distinct observation to be made from the plot is that the six phages Carlos, Rosalind, GIW2016, ABAtENZ, AJG77 and ApiWellbeing are almost identical to each other and clearly distinct from the first three, i.e. Callan, Dash and Lilo. In the smaller cluster of three genomes, Callan and Dash seem very similar, whereas

[15] www.mdpi.com/1999-4915/17/2/137
[16] www.ncbi.nlm.nih.gov/sites/batchentrez

Lilo differs somewhat from the other two. According to the article, the PNI values for Lilo compared to Callan and Dash were approximately 74% and 80%, respectively (Kok et al., 2025), which in the plot leads to slightly fragmented, but still clearly identifiable diagonals. Further, it can be noticed that all nine genomes seem to share three regions of a few thousand bases in length, which appears as repeated patterns of short diagonals occurring in all pairwise comparisons.

a

b

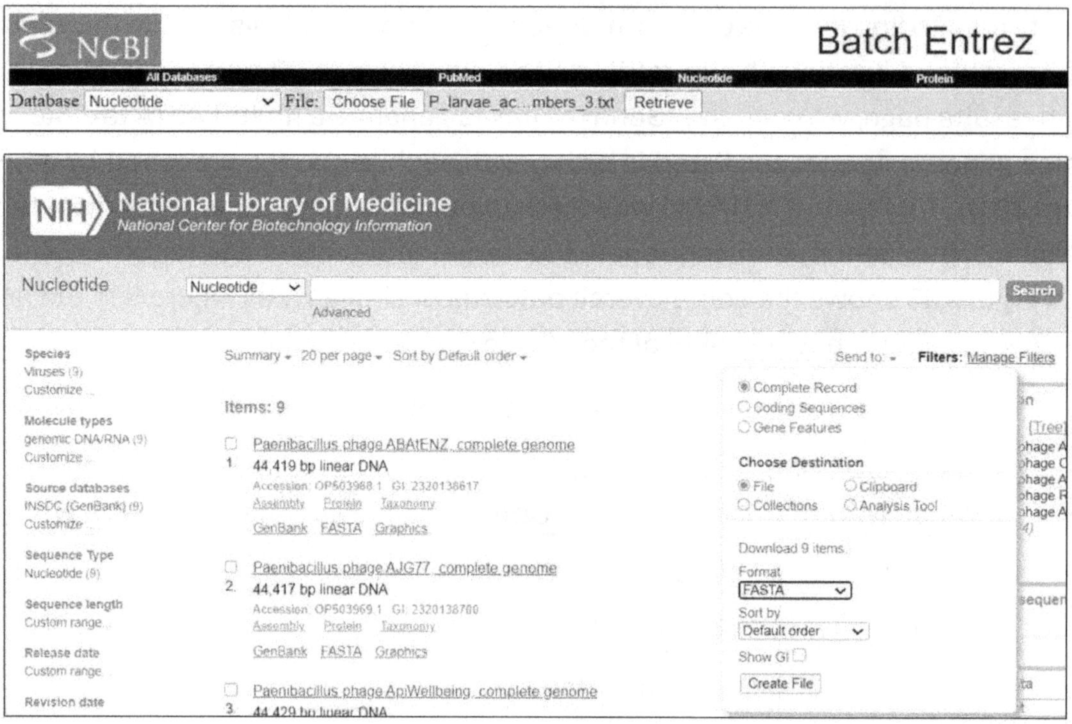

Figure 39. Using the Batch Entrez interface to find a set of sequence entries. In the upper panel (a), the file containing the list of accession numbers is uploaded and then the entries are retrieved. The lower panel (b) shows the search results and how a multi-FASTA file containing all retrieved sequences (nine in this case) can be created and downloaded.

Upon close inspection of the dot plot in **Figure 38**, we can see a synteny break close to the middle of the first two genomes, from the phages Callan and Dash. **Figure 40** shows a dot plot between these two sequences only, where the synteny break is much more clearly visible. The break indicates a genomic region that is present in the Dash genome but absent in the Callan genome. In their article, Kok and co-workers identified a gene coding for a toxin, Plx1, which contributes to pathogen virulence. The Plx1 gene is located in the divergent region and therefore present in Dash but not in Callan. The Plx1 toxin contains an A-subunit that is thought to induce apoptosis in eukaryotic cells and a B-unit that helps mediate entry into the host cell (Kok et al., 2025). Although this is not the only source of pathogenicity of *P. larvae* phages, the presence or absence of this gene is clearly a factor to consider when choosing a phage for the development of phage therapy.

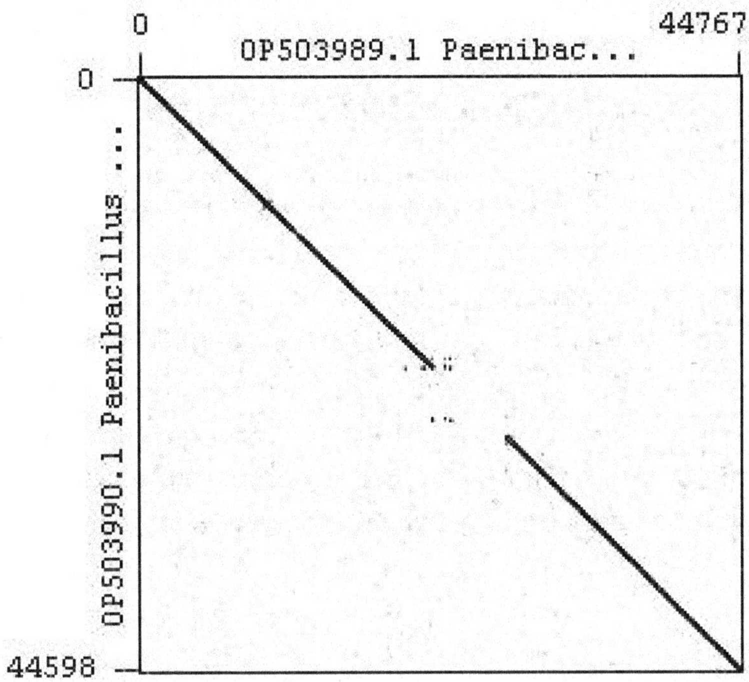

Figure 40. Gepard dot plot of the genomes of the phages Callan and Dash. Generated with automated parameter settings (window length 10).

Application example 2: Identifying gene duplication events

We will here reproduce some of the results from a study that aimed to gain new insights into how the diversity of haemoglobin genes has evolved in fish genomes as an adaptation to the large variations in oxygen concentration in different aquatic environments (Omelchenko et al., 2024). Haemoglobin serves the role of carrying oxygen from the lungs to body tissues, and to carry carbon dioxide to the lungs, and is thereby crucial for aerobic respiration. Teleost fishes, which make up the vast majority of all extant fish species, show an unusually high variation and high number of haemoglobin genes compared to other vertebrates. This can likely be explained by the variation in oxygen concentrations being much larger in aquatic environments than on land (Omelchenko et al., 2024). As an example, the authors mention the cichlid fishes living in deep tropical lakes, for example Lake Malawi, where the steep oxygen gradient represents one of the most important axes of differentiation between species.

Haemoglobin is a protein complex, i.e. it consists of subunits. For example, the human haemoglobin A, which makes up the vast majority of all haemoglobin in the red blood cells of adults, is a tetramer $\alpha_2\beta_2$ consisting of two haemoglobin alpha and two haemoglobin beta subunits. Each of these subunits can be encoded for by a different number of genes in various species. The human genome contains two alpha subunit genes, *HBA1*[17] and *HBA2*[18], with identical coding sequences, but also three additional genes for the alpha-like haemoglobins *zeta*, *theta*-1 and *mu*, which are expressed in early developmental stages. In total, therefore, the cluster of Hbα genes in the human genome, located on chromosome six, contains five protein-coding Hbα genes. Similarly, human chromosome 11 contains a cluster of five Hbβ genes, with *HBB* coding for the beta subunit protein present in the haemoglobin A tetramer, and four related beta-like genes.

In contrast to human and other tetrapods, teleost fish do not have their haemoglobin genes organised in separate Hbα and Hbβ clusters, but in two genomic clusters where each cluster contains mixed Hbα and Hbβ subunit genes (Omelchenko et al., 2024). In the Omelchenko et al. study, a remarkably high variation in the number of haemoglobin genes was observed in the 17 teleost fish genomes included in the comparison. The number of genes ranged from five in the Japanese pufferfish (four Hbα subunit genes and one Hbβ subunit gene) to 43 in Atlantic salmon (20 Hbα subunit and 23 Hbβ subunit genes). Within the cichlid family, from

---

[17] www.ncbi.nlm.nih.gov/gene/3039
[18] www.ncbi.nlm.nih.gov/gene/3040

which six genomes were included in the study, the number of haemoglobin genes varied from 15 to 32.

The Gepard dot plot in Figure 3 in the Omelchenko et al. article is an attempt to illustrate how the large number of subunit genes has evolved in one of the haemoglobin gene clusters of Nile tilapia (*Oreochromis niloticus*), which had the largest number of subunit genes among the cichlid fish included in the study. The plot indicates three regions of tandem duplications. In Figure 3 in their article, the authors have marked these regions in different colours. Using the reproduced plot, shown in **Figure 41**, we will here refer to the regions we discuss using the letters *a*-*e*. Close to the upper left corner, the small repeat region (*a*) indicates a recent duplication where the Hbβ and Hbα gene pair, named Mn1 and Mn2, respectively, in Figure 3 in the article, is repeated as the corresponding pair Mn3 and Mn4. That this duplication happened recently can be inferred from observing that the corresponding diagonals, to the left and right of the main diagonal, are unbroken, indicating a very high degree of sequence identity.

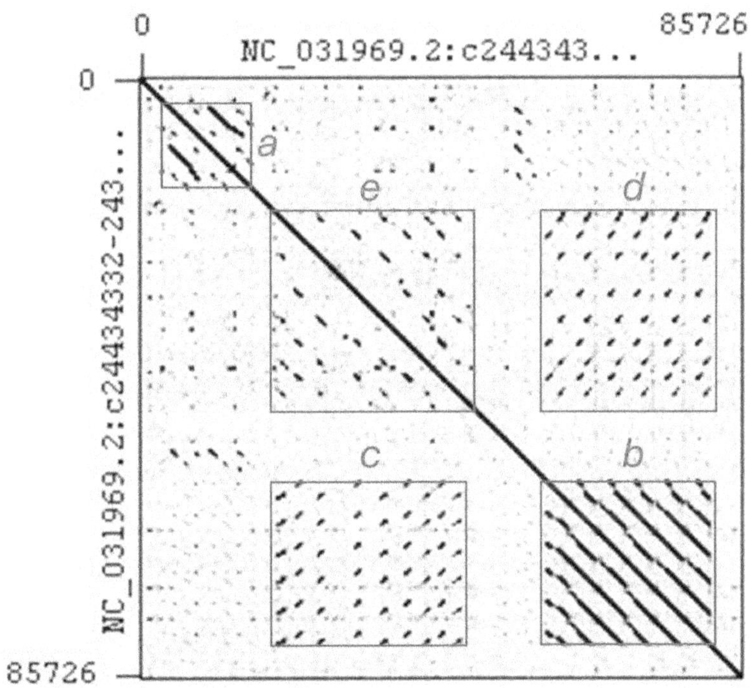

Figure 41. Gepard dot plot of 86 kb region of the *O. niloticus* genome containing the Hb gene cluster. Letters *a*-*e* indicate tandem repeats discussed in the text.

The second indicated region (*b*) shows the result of successive duplications having resulted in six instances of the Hbβ and Hbα gene pair, numbered as Mn19 to Mn29 in Figure 3 in the article (Omelchenko et al., 2024). These duplications also seem to be recent events since the corresponding dot plot diagonals are again almost without any gaps. Based on the multi-species genome comparisons in the article, the authors suggest that this expansion of Hb genes most likely started in the oreochromine ancestor, since it is not observed in other teleost fish genomes, and that it continued exclusively in *Oreochromis*.

The two regions *c* and *d* in **Figure 41** indicate that the original duplication leading to gene cluster *b* appeared by translocation and inversion, since the orientation of genes in region *b* is in the opposite direction compared to the older gene cluster *e*. The *e* cluster contains four Hbα and Hbβ pairs, followed by a single Hbα gene, i.e. a total of nine Hb genes, Mn6 to Mn14. This cluster is older and shared among other cichlid fish. The fragmented and unclear diagonals representing the tandem repeats in this region reflect the larger number of accumulated mutations in this region.

The reproduced figure (i.e. **Figure 41**) was generated as described in the following. It was first noted that the caption of Figure 3 in the article (Omelchenko et al., 2024) referred to the accession number NC_031969.2 as the source for the sequence data. This is an NCBI reference sequence containing the full *O. niloticus* genome. The plot, however, contains only the region with coordinates 24,434,332 – 24,348,606, i.e. an approximately 86 kb region containing the analysed Hb cluster. To download only this part of the sequence, we set the "Selected region" parameters accordingly in the "Change region shown" menu on the FASTA format web page for the NC_031969.2 entry in the NCBI nucleotide sequence database (**Figure 42**). It would be possible to download this sequence region and make the dot plot, but we would then get a plot that would be a mirror image of the one from the article. In order to get a dot plot that is easily compared to the one in the article, we need to also set the "Display options" parameter in the "Customize view" menu to the value "Show reverse complement". With these settings, we download the 86 kb FASTA sequence to a text file, which is then uploaded into the Gepard tool and plotted against itself with default settings.

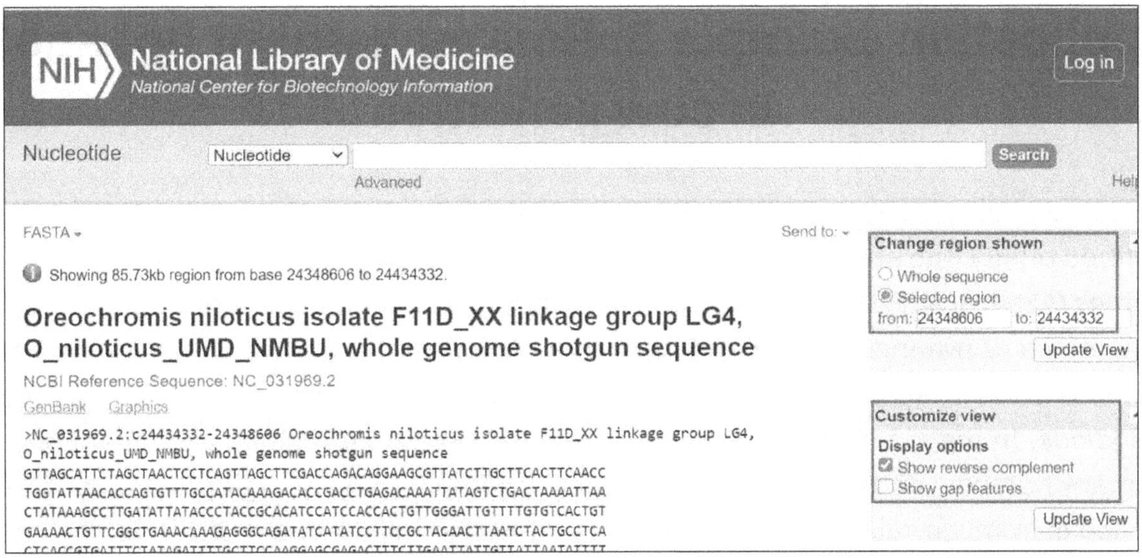

Figure 42. Downloading the reverse complement sequence of the selected 86 kb region of the O. Niloticus genome, Genbank accession number NC_031969.

The article contains several additional dot plots that could be of interest to reproduce in order to get used to the dot plot technique. In Figure 5a (Omelchenko et al., 2024) the Hb region of another cichlid, *Pungu maclareni*, is plotted both against itself and against *O. niloticus*. In Figure 5b, four other cichlid genome regions are plotted, each against itself. The comparison of these four plots shows that the older duplication region, which was marked as region *e* in our plot in **Figure 41**, is shared by African cichlids only, and that the more recent duplications (such as those marked with *a* and *b*) seem to have occurred independently in different cichlid species.

Application example 3: Rearrangement hot spots in cow genomes

Yet another research example where the Gepard dot plot tool was used will here be described briefly. The study concerns the identification of rearrangement hot spots in cattle genomes (Azam et al., 2025). The authors sequenced, assembled, annotated and compared the genomes of five Indian breeds of cattle. The five "new" genomes were also compared with the previously assembled and annotated genome representing the Brahman breed of *Bos indicus*. In the synteny analysis, many perfectly collinear regions were identified, but also many hotspots of rearrangement. The rearrangement hotspot regions were found to be enriched in immune-related genes and genes related to milk production. Thus, the regions with many genome rearrangements were related to complex traits relevant to breeding (Azam et al., 2025).

Figure 7 in the article by Azam et al. (2025) shows an 800 kbp long region in chromosome 23 of the Brahman genome plotted against itself. This region contains the major histocompatibility complex (MHC), and synteny analysis across genomes showed that the breeds differed in their gene copy numbers in this region. The synteny plot in Figure 7b shows how the number of immune-related genes (shown in green) differed a lot between the breeds, while the region was flanked by orthologous non-immune-related genes that were present in all compared genomes. Looking at the dot plot in Figure 7a (Azam et al., 2025), we can see clear traces of gene duplication events within this region in the form of several short diagonals in parallel to the main northwest-southeast diagonal. Each pair of short diagonals to the left and right of this diagonal represents one probable duplication event. Also, the degree of clearness of each diagonal, i.e. whether it appears as a solid, broken or dotted line, can be taken as an indication of whether the duplication occurred recently or a long time ago.

To reproduce the figure from the article, we need to first, as usual, find the sequence data. In this case it is stated in the article that the Brahman genome assembly UOA_Brahman_1 from NCBI was used. When searching the NCBI nucleotide database for this identifier, we found a Genbank entry with accession number PUFS00000000.2. This is not a sequence entry but a master entry for data from a genome sequencing project. Such entries provide general information about the biological background and the sequencing project, whereas the sequence data is linked to in the WGS field. In this case, the relevant text line in the master entry looks as shown below and also has a hyperlink:

WGS PUFS02000001-PUFS02001251

When following this link, we are redirected to the NCBI Sequence Set Browser, which is a web interface for browsing through all data related to a particular sequencing project. The link should lead to the following URL:

https://www.ncbi.nlm.nih.gov/Traces/wgs/PUFS02

In the browser we now see a table listing all contigs from the assembly that was done in the Brahman genome sequencing project. Scrolling down to entry 23 in this table and following the link to the FASTA sequence entry, we find that it contains a contig covering the complete chromosome 23, with NCBI accession number PUFS02000023.1. From this sequence entry, we need to download the 800 kbp segment with coordinates 29100000-29900000, so these coordinates should be given as the start and end point for the "Selected region" in the "Change region shown" menu.

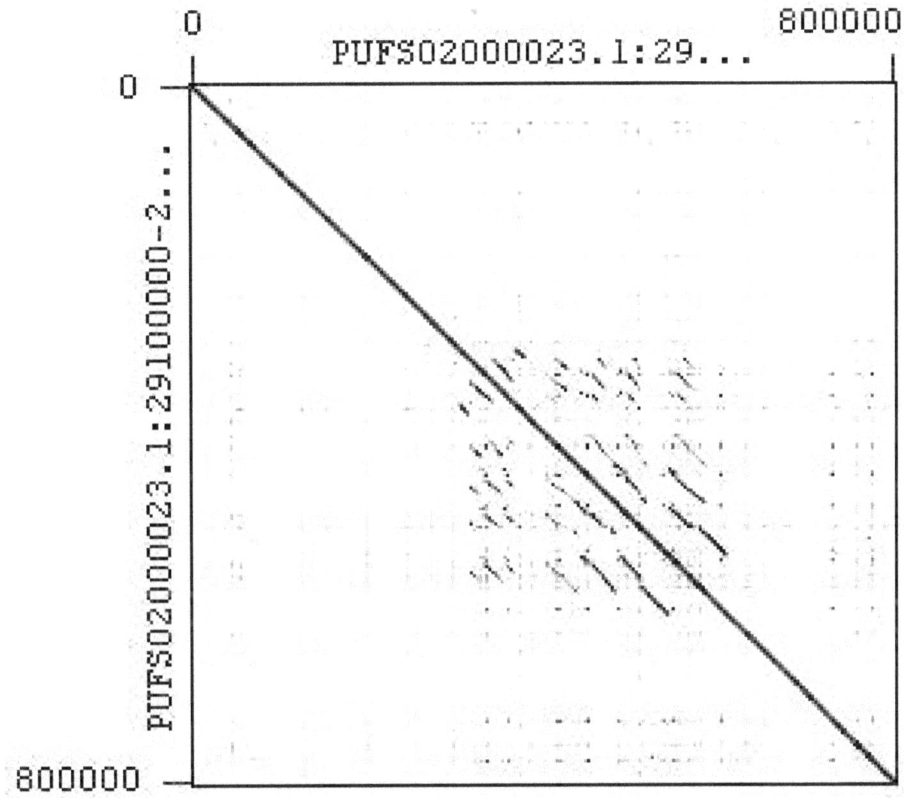

Figure 43. An 800 kbp long region of the genome of the Brahman breed of *Bos indicus* plotted against itself, using Gepard, revealing a "hot spot" for genome rearrangements.

When the selected sequence has been downloaded to a text file, we upload this file into Gepard and plot the sequence against itself with default parameter settings. After some tuning of the display settings (you may have to adjust the settings for "Lower color limit" and "Upper color limit" to filter out irrelevant matches), the result should look as in **Figure 43,** i.e. as a reproduction of Figure 7a of the article (Azam et al., 2025).

Application example 4: Structural differences between rice genomes

We will now discuss examples where the D-GENIES tool was used to compare longer sequences. In the first example, the genome of the cultivated rice, *Oryza sativa*, was compared to the genomes of two of its wild progenitors, *Oryza rufipogon* and *Oryza meridionalis* (Abdullah et al., 2025). These wild rice species are native to tropical Asia and Northern Australia, respectively, and the authors of the article state that they can be seen as genetic reservoirs for potential breeding purposes and adaptation to changing environments. Among these two progenitors, it seems that *O. rufipogon* is the closest relative to domesticated rice, and that there has been significant gene flow between wild and domesticated rice species in Asia for thousands of years. In contrast, the wild rice in Northern Australia has been isolated from the process of domestication and may therefore have preserved more pre-domestication traits (Abdullah et al., 2025).

To gain more insight into the genetic traits of the progenitor species and the dynamics of rice speciation, the authors performed full genome sequencing of *O. rufipogon* and *O. meridionalis* and carried out a comprehensive comparison between these genomes and previously published genomes from the *O. sativa* subspecies Nipponbare and Indica. In this comparison, it was noticed that a 6 Mb inversion on chromosome 6 was present in the Australian wild rice, *O. rufipogon*, when compared with the *O. sativa* Japonica genome. It was further noticed that this inversion was also present when the Asian wild rice, *O. meridionalis* was compared to Japonica. However, when each of the two species of wild rice were compared with *O. sativa* Indica, the inversion was absent. From these observations, the authors concluded that the inversion on chromosome 6 represents a critical divergence between the subspecies Japonica and Indica. The presence and absence of the inversion is visible in the full genome synteny maps in Figures 3a and 3b in the article. It is also studied in more detail in the dot plots of chromosome 6 in Figure 3c (Abdullah et al., 2025). **Figure 44** shows a reproduction of the *O. rufipogon* versus Japonica (Nipponbare) chromosome 6 dot plot, which is the upper left of the four dot plots shown in Figure 3c in the article by Abdullah et al. (2025).

As preparation for creating this dot plot, we downloaded the FASTA file for entry CP150379.1 from the NCBI nucleotides sequence database. This entry contains the sequence of *O. rufipogon* chromosome 6. To find the corresponding sequence from Japonica, we used the genome database Ensembl Plants[19] (Yates et al., 2022) to find the reference genome

[19] plants.ensembl.org

assembly for *Oryza sativa* Japonica. We there found the assembly under the name IRGSP-1.0. For the sequence data, Ensembl Plants links to the ENA[20] (European Nucleotide Archive) database (O'Cathail et al., 2025), where the set of sequence files can be found in the entry with accession number GCA_001433935. From the main page of this ENA entry, there is a link to "Chromosomes", leading to a page with a listing of all chromosomes in the Japonica reference genome. We here found that the entry for chromosome 6 has accession number AP014962. Finally, on the web page[21] for the entry AP014962, there is a link to download the sequence of chromosome 6 in FASTA format.

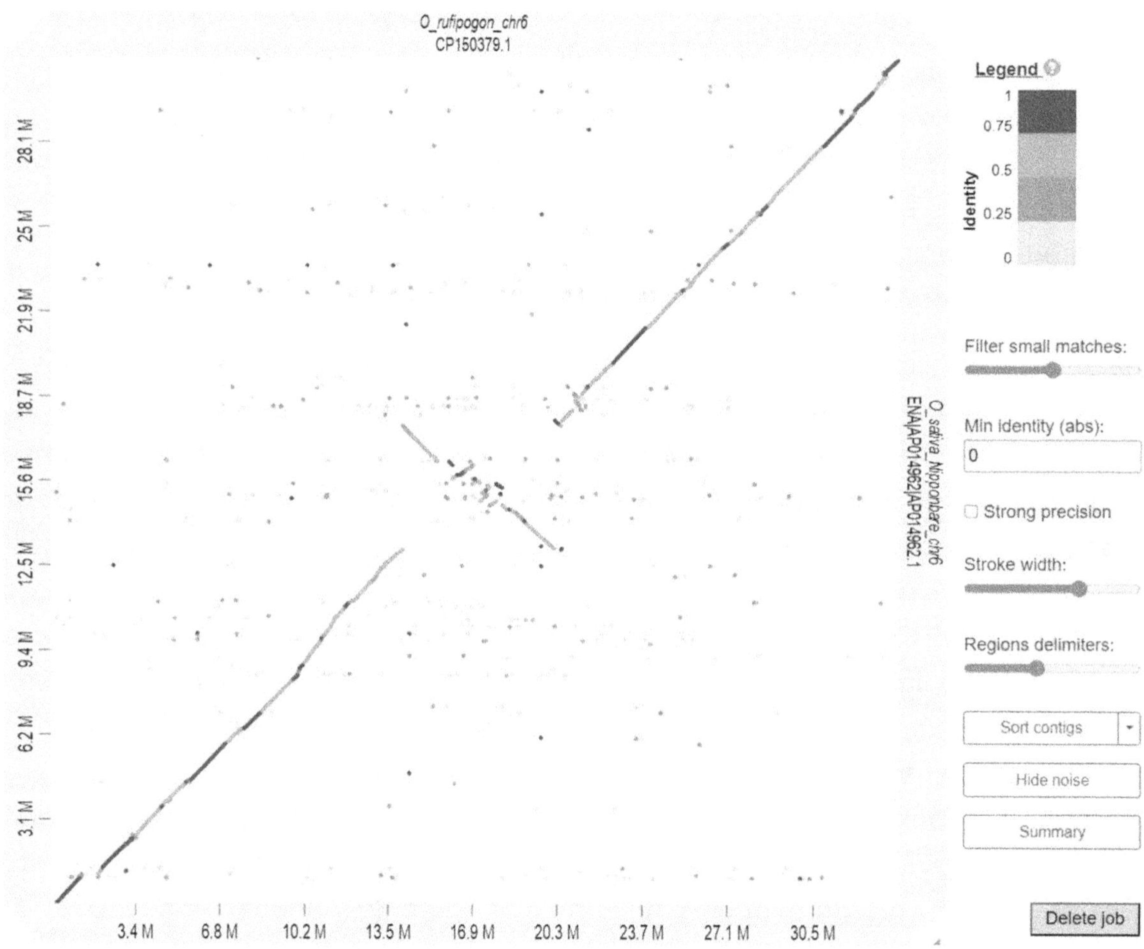

Figure 44. D-GENIES dot plot comparing chromosome 6 of the Asian wild rice *Oryza rufipogon* and the cultivated *Oryza sativa* subspecies Japonica (Nipponbare).

For the plot in **Figure 44**, we used the online version of D-GENIES. This is done by visiting the home page[22] of the project and following the link "Run". On this page we only need to provide an email address and upload the two sequences. The plot was created using default

[20] www.ebi.ac.uk/ena

[21] www.ebi.ac.uk/ena/browser/view/AP014962.1

[22] dgenies.toulouse.inra.fr

settings. Despite the sequences being as long as approximately 33 Mb each, the runtime for generating the plot was only 7 seconds. For comparison, when attempting to make the corresponding plot with Gepard, the software will produce a message saying that the sequences are too long and that there is insufficient memory for generating the suffix array.

When the figure had been generated, we changed the settings for "Stroke width" and "Filter small matches" to values higher than the default settings to make the visualisation of relevant matches clearer[23]. D-GENIES uses a colour coding of matches to show the level of sequence identity. In **Figure 44**, we can see how the percent identity along the main diagonal is either 50-75% (light green) or 75-100% (dark green) for most of the length of the diagonal. The orange and yellow colouring of shorter matches indicate lower levels of percent identity. When inspecting the inversion in the central part of the plot in more detail, we can see that the middle part of the inverted segment is quite fragmented. It also seems that additional inversion events have led to parts of the segment reverting to the original orientation.

The supplementary files of the article by Abdullah et al. (2025) contain, among other material, pairwise dot plot comparisons of full genomes. The rice genomes consist of 12 chromosomes of approximately equal length; thus, the full genome size is approximately 500 Mb. Although this is a relatively compact genome compared to other cereals, it is still far above the length that can be addressed by dot plot tools based on suffix arrays, such as Gepard. As a test of the computing time, we downloaded the full genomes of *O. rufipogon* and Japonica and generated a dot plot through the D-GENIES online tool. Runtime in this case was approximately 1 minute, generating the plot shown in **Figure 45**.

In this case, we downloaded the 12 chromosome sequences *for O. rufipogon* by first finding the reference genome entry in the NCBI genome database, with accession number GCA_037997075. From the web page for this entry, we used the link "View Genbank sequences", which led to a Genbank search hit list including the 12 entries CP150374, CP150375, ..., CP150385. From the page with the search hits, we downloaded all 12 sequences into a single multi-FASTA file using the "Send to" link. For Japonica, we started from the previously mentioned ENA entry for the genome assembly (GCA_001433935) and, under the heading "Download", used the link "All Seq Fasta", downloaded the corresponding multi-FASTA file for Japonica. The two multi-FASTA files were then uploaded into D-GENIES, which was again run with default settings.

[23] For explanations of additional parameter settings on the results page, see the documentation at dgenies.toulouse.inra.fr/documentation/result

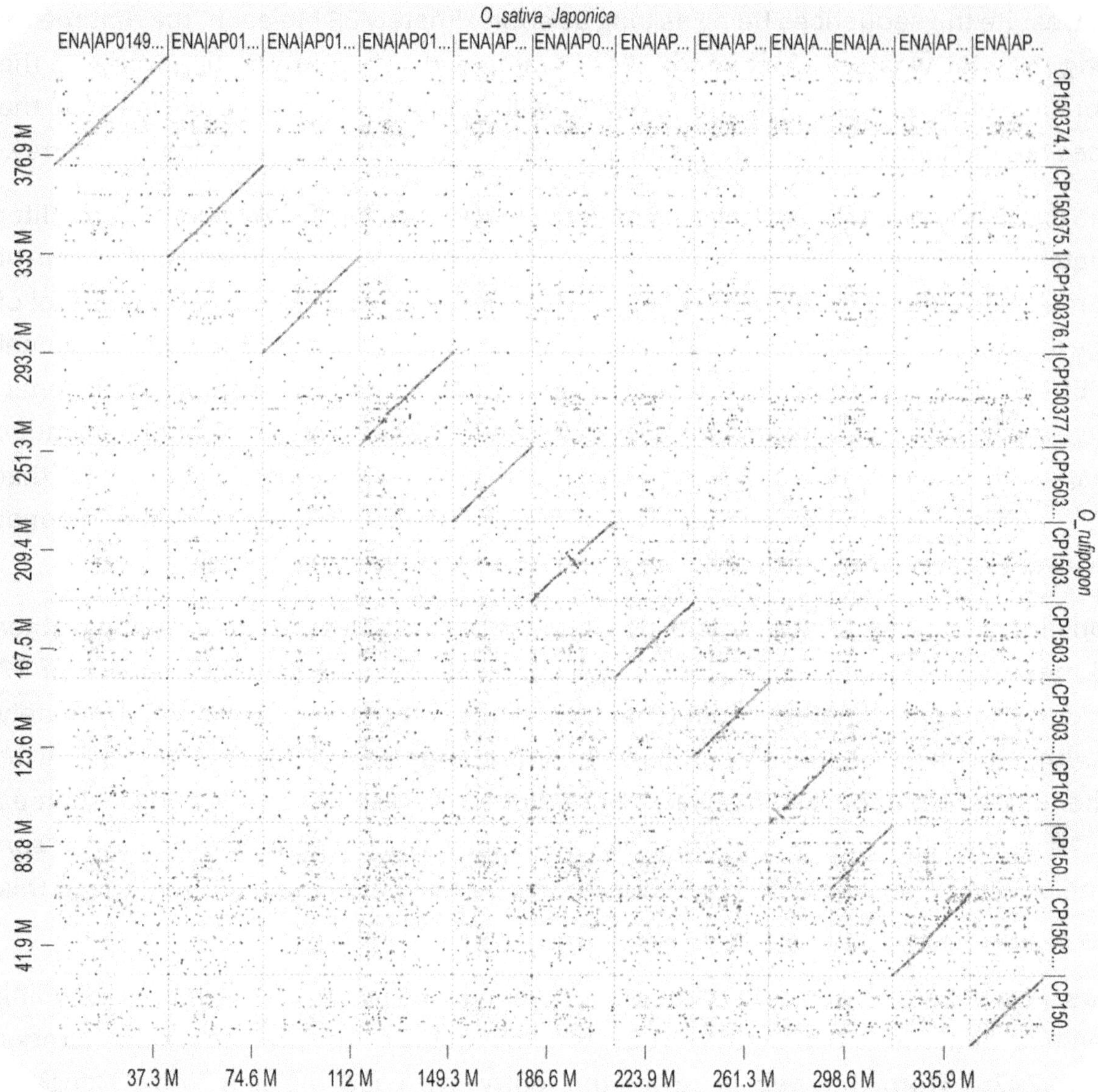

Figure 45. D-GENIES dot plot comparing the full genomes of Asian wild rice *Oryza rufipogon* and the cultivated *Oryza sativa* subspecies Japonica (Nipponbare), consisting of 12 chromosomes each.

As can be seen in **Figure 45**, the two sets of sequences are in opposite orientations. One way to adjust for this would be to download each chromosome from NCBI separately and use the reverse complement option for each downloaded sequence. As alternative methods, there are two ways of reorienting the sequences in the D-GENIES results page. The first method is to select one target-query contig[24] pair at a time and reverse the orientation of the query sequence. The "target" is the horizontal sequence, i.e. in our case *O. rufipogon*, and the

[24] A "contig" is a contiguous sequence, i.e. a sequence without any gaps. In the context of the example discussed in this chapter, it means that each chromosome has been assembled as a single contiguous sequence, rather than a set of shorter sequences corresponding to separate parts of the chromosome with gaps in between.

"query" is the vertical sequence, i.e. *O. sativa* Japonica. A contig pair in this case is a complete chromosome sequence from *O. rufipogon* matched with a chromosome sequence from Japonica. For example, in the upper left corner of our plot, we see the match between the two organisms' sequences for chromosome 1. To reverse this match, we can right-click on it and then select "Reverse query" in the menu that pops up. In **Figure 46**, we have first reversed the query for chromosome 1, then right-clicked on the contig pair for chromosome 2 so that the menu with the "Reverse query" option is shown. Continuing in this way, reversing one of the 12 query contigs at a time, we would get the full genome match shown as a diagonal from the northwest to the southeast.

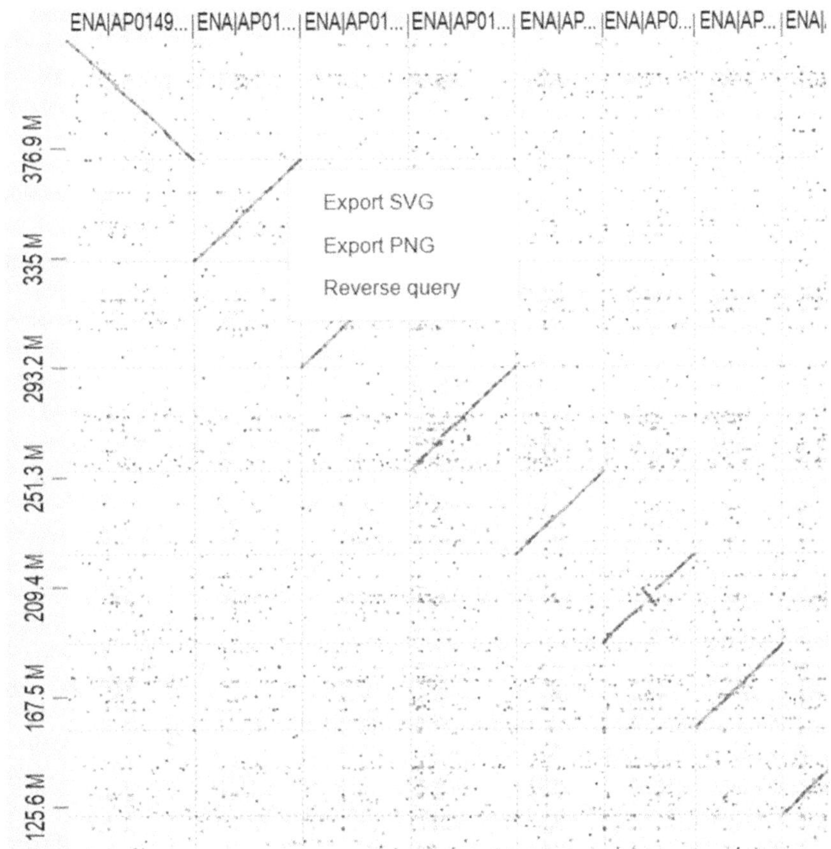

Figure 46. Using the "Reverse query" function on the D-GENIES results page to reverse the orientation of matches, chromosome by chromosome.

A second way of adjusting the orientation using parameter settings on the D-GENIES results page is to use the "Sort contigs" button. This has the effect that the contigs of the query will be sorted according to the reference. The D-GENIES documentation states that for each contig in the query, the program searches for the region which has the "biggest" matches with the query and stores these coordinates. It then sorts the contigs according to their coordinates. In our example, comparing the two rice genomes, the chromosomes were already sorted in the same order, but the contigs were in opposite orientation. In this case, when using the "Sort contigs" function, the orientation of the query sequences will be reversed, producing the plot shown in **Figure 47.** Here, after the sorting has been done, the label on the "Sort contigs" button has changed to "Undo sorting". Note that if individual query contigs are first reversed, as shown in **Figure 46**, and you then attempt to use the sorting function to reproduce **Figure 47**, you will first have to use the "Reset" function to revert back to the original plot.

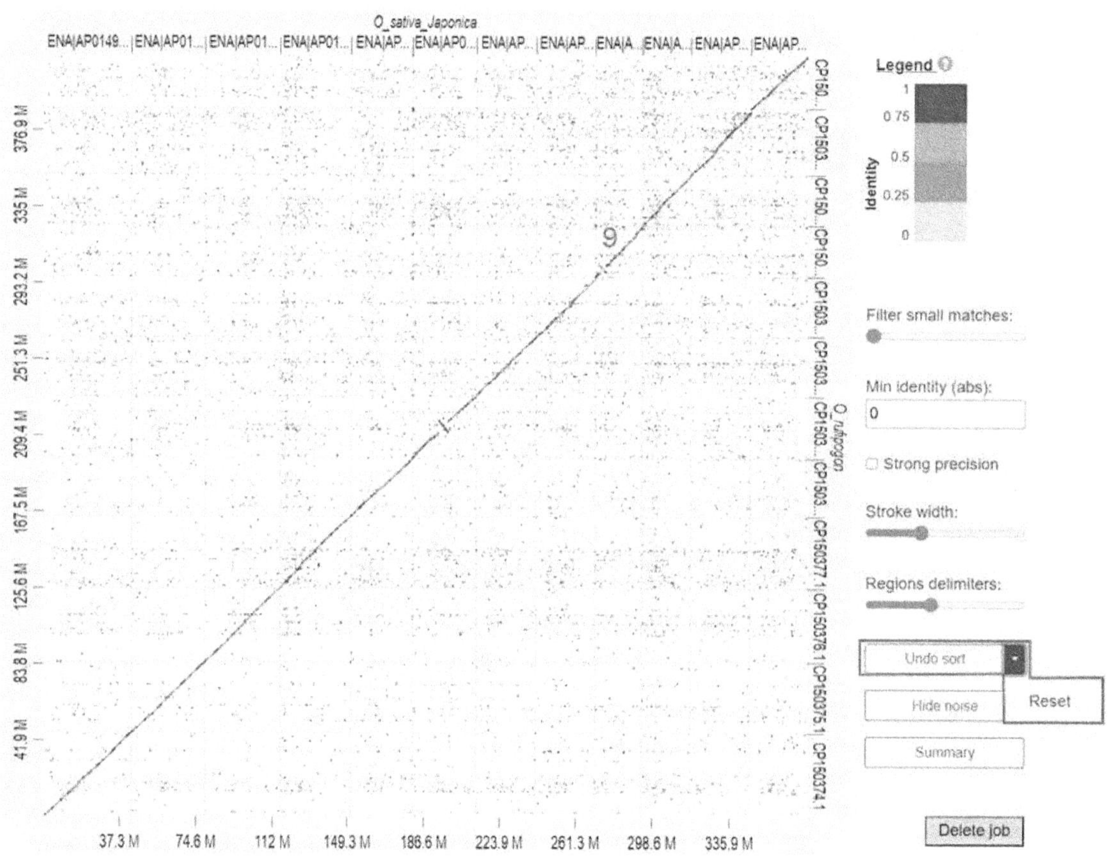

Figure 47. Using the "Sort contigs" function produces a plot where the genomes are displayed in the same orientation. The box marks the button used to sort contigs and to undo the sorting. This button also has a "Reset" option, which is used to revert back to the original plot. The number 9 in the plot indicates the match between chromosome 9 of each species, which is the dot plot zoomed in on in **Figure 48**.

Another very useful functionality on the D-GENIES output page is the possibility to zoom in on individual contig pairs by clicking in the graph. In the plot shown in **Figure 47** we had increased the setting for "Region delimiters". The lines showing the beginning and end of each contig then become more visible in the plot area, forming a grid pattern, where every square corresponds to an individual plot between two contigs. It thus becomes apparent that what we are looking at is not actually a single dot plot, but rather a set of $n \times m$ dot plots, with *n* and *m* representing the number of contigs in the Japonica and *O. rufipogon* data sets, respectively. In order to inspect a single plot, we can simply click on the corresponding square in the visualisation. Noticing that the plot for chromosome 9, marked with the digit 9 in **Figure 47**, seems to contain an inversion that is almost as long as the one observed in chromosome 6, we therefore clicked in the square marked with 9. This produced the plot shown in **Figure 48**, where we can see a large inversion near the beginning of the sequences, as well as other smaller rearrangements. In this simple way, the overall genome plot can be easily navigated by clicking on the individual dot plots representing single chromosomes.

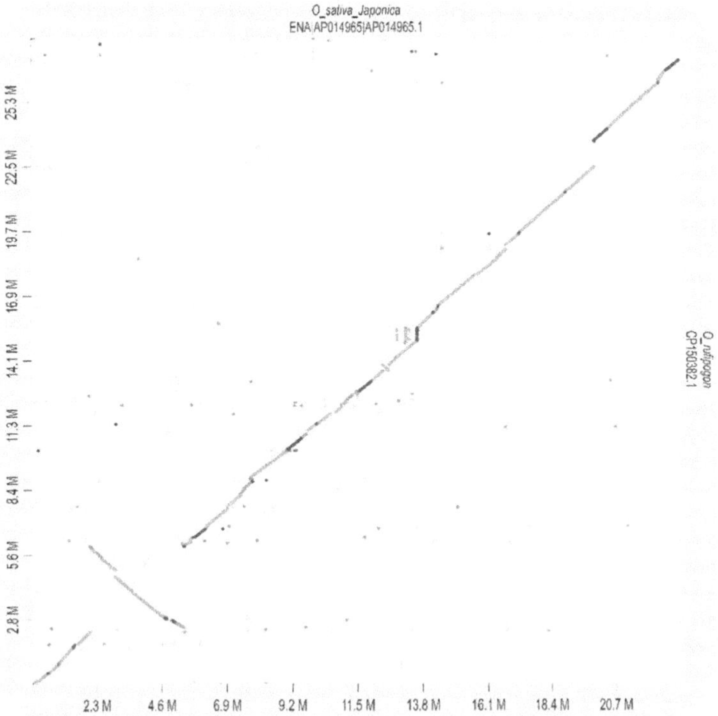

Figure 48. Zooming in on a single chromosome in the genome comparison between rice species highlights a large inversion and several smaller rearrangements in chromosome 9.

Application example 5: Measuring the overall genome sequence similarity

The final application example demonstrates that the use of minimizers enables fast construction of dot plots for large eukaryote genomes. In the study that the example is based on, the reference genome of the Spanish toothcarp was established (López-Solano et al., 2025). The Spanish toothcarp, *Aphanius iberus*, is native to marshy waters along the Mediterranean coast of Spain. In the article the authors explain that this fish is a euryhaline species, meaning that it has great capabilities for adaptation to different levels of salinity and temperature. It is therefore of interest as a potential model species for understanding the genetic mechanisms underlying this adaptability.

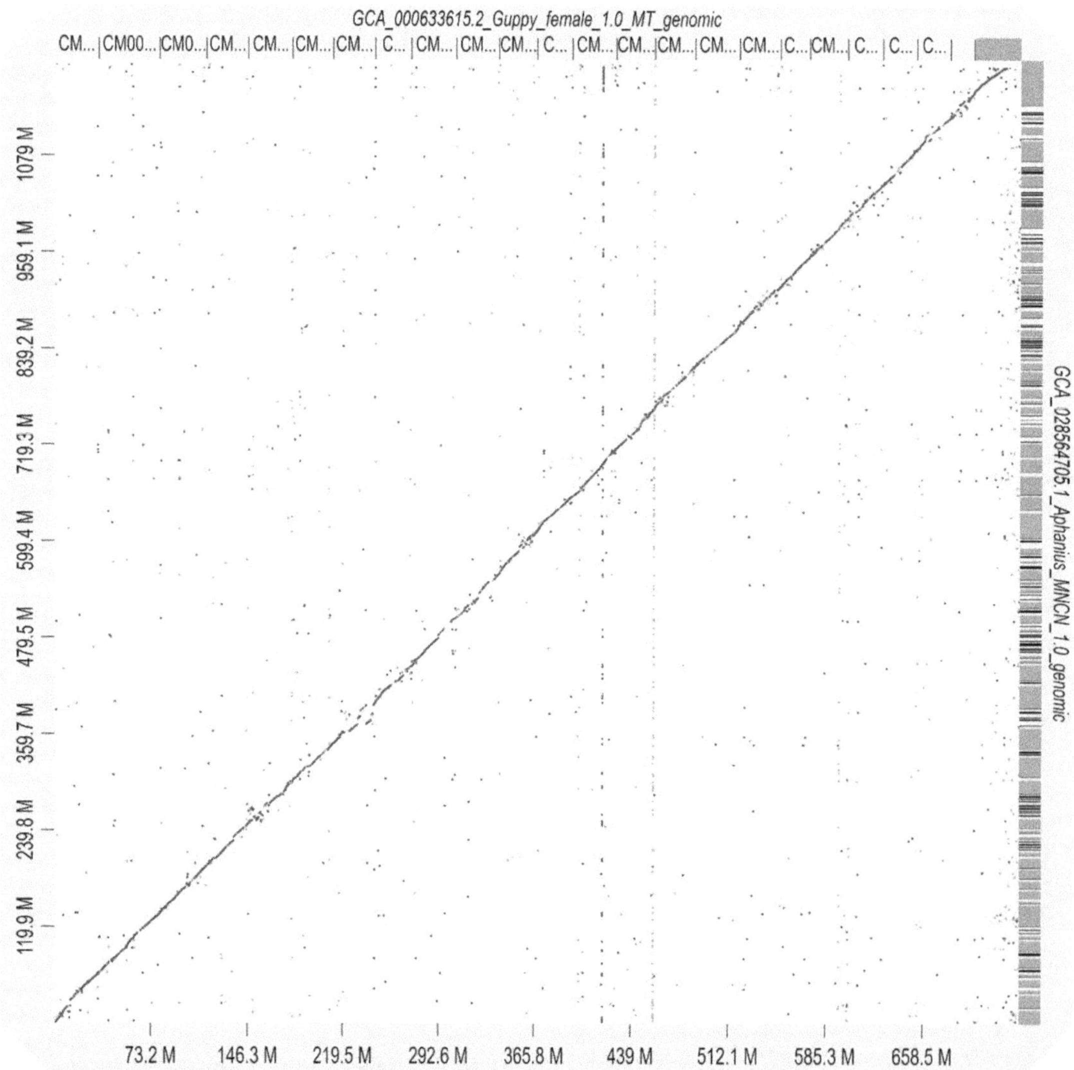

Figure 49. D-GENIES comparison of assemblies of the genomes of the two fish species *Aphanius iberus* and *Poecilia reticulata*.

The authors presented a full genome assembly of *A. iberus* and compared it to the genome of *Poecilia reticulata*, which is a closely related species with a well-characterised genome. In Figure 4 of the article (López-Solano et al., 2025), a dot plot comparison of the two genomes, generated with D-GENIES, was shown. In **Figure 49** we show our reproduced version of this dot plot. Given the large sequence sizes, with the *A. iberus* genome containing approximately 1.2 billion base pairs, the execution time for generating the plot was just under two minutes. This can be considered remarkably fast, given a genome length of approximately a third of that of the human genome.

To generate this dot plot, we first searched for the *P. reticulata* reference assembly entry in the NCBI nucleotide database, with accession number GCF_000633615. From the web page for this database entry, we used the "Download" link and started the download with the setting "All" for "Select file source" and "Genome sequences (FASTA)" for "Select file types" (**Figure 50**). We thus obtained a compressed archive, containing, among other files, a text file containing all assembled scaffolds in multi-FASTA format. Using the same steps, we also downloaded the sequence data for *A. iberus* from the NCBI entry with accession number GCA_028564705. Before uploading the two sequence files to the D-GENIES server, it should be noted that the tool accepts compressed files in gzip format as input. This may be useful in this case, since the two uncompressed multi-FASTA files together make up a size of 1.7 Gb.

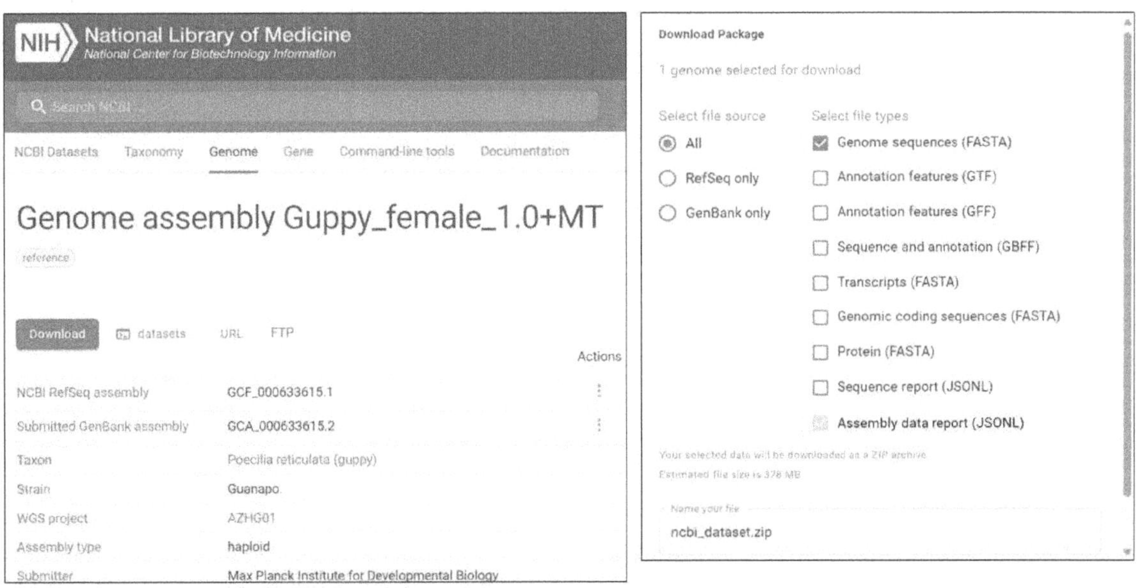

Figure 50. Downloading the set of contigs for the *P. reticulata* assembly. Left panel: The NCBI database entry holding the data set (accession number GCF_000633615). Right panel: Selecting the suitable file types for the download.

A slight complication in the comparison of these genomes is that the reads from the sequencing have been assembled at different levels. The assembly statistics in the database entries show that *P. reticulata* contigs have been assembled into 23 complete chromosomes, whereas the *A. iberus* contigs have been assembled only at scaffold level (**Figure 51**). A scaffold is a set of contigs that have been linked together by bridging the gaps between them using information from paired reads (Pop et al., 2004). Although scaffolds are, obviously, longer than contigs, they do not necessarily cover the full lengths of chromosomes. As seen in the assembly statistics, both datasets include approximately three thousand scaffolds (3028 for *P. reticulata* and 3054 for *A. iberus*) and they also have relatively similar N50 values[25] of 5.3 and 1.7 Mb, respectively, indicating that the two assemblies are of comparable overall quality. However, the assemblies differ in the important aspect that in one case scaffolds were linked together to represent chromosomes, whereas in the other case they were not.

---

[25] N50 is a measure of the degree of fragmentation of an assembly, where a higher value means less fragmentation and therefore a better assembly. The N50 value is found by first sorting all scaffolds (or contigs) by length, from longest to shortest, and then summing up their lengths, starting with the longest one. When 50% of the total length of all scaffolds has been reached, the length of the last scaffold included in the sum is reported as the N50 value. For our example data, this means that 50% of the total length of the *P. reticulata* assembly is made of scaffolds of length 5.3 Mb or longer. For details on N50 and similar assembly quality measures, see (Castro & Ng, 2017).

Assembly statistics

	RefSeq
Genome size	731.6 Mb
Total ungapped length	664.6 Mb
Gaps between scaffolds	261
Number of chromosomes	23
Number of organelles	1
Number of scaffolds	3,028
Scaffold N50	5.3 Mb
Scaffold L50	43
Number of contigs	40,143
Contig N50	41.9 kb
Contig L50	4,423
GC percent	39.5
Genome coverage	110.0x
Assembly level	Chromosome

Assembly statistics

	GenBank
Genome size	1.2 Gb
Total ungapped length	1.2 Gb
Number of scaffolds	3,025
Scaffold N50	1.7 Mb
Scaffold L50	201
Number of contigs	3,054
Contig N50	1.6 Mb
Contig L50	205
GC percent	39
Genome coverage	95.0x
Assembly level	Scaffold

Figure 51. Assembly statistics for the *P. reticulata* (left panel) and *A. iberus* (right panel) genome sequencing data sets.

As a consequence of the different assembly levels, the *x*-axis on the upper edge of the plot (**Figure 49**) shows 23 delimited regions of the *P. reticulata* genome, corresponding to the 23 chromosomes. In comparison, when looking at the *y*-axis on the right-hand side of the plot, we see the *A. iberus* genome delimited into a very large number of small regions, representing the 3054 scaffolds. This is a complication when trying to zoom in to look at how individual pairs of chromosomes match between the genomes, since each chromosome in the *A. iberus* sequence is represented by a large number of scaffolds in the plot. As we will see a little bit later on, the zoom function will in this case show a plot where a chromosome is compared with a scaffold.

In the article, the authors comment on the plot, stating that the "*results based on a* 0.75 *identity level suggest that these two genomes exhibited high whole-genome scale similarity*" (López-Solano et al., 2025). As explained in the caption of Figure 4 in the article, their plot showed only the matches with sequence identity over 75%, and the plot looks convincing as support of the statement. However, it is somewhat open to interpretation what the authors meant with "high whole-genome scale similarity", and it should perhaps not be taken to mean that the average overall sequence identity is above 75%, but rather that the genomes show high structural similarity, with very few large-scale rearrangements. Be that as it may, the authors of the D-GENIES software suggest that overall sequence similarity should be estimated using the identity summary function. The identity summary shows the sums of projections of matches on the reference sequence per similarity category (where the categories are no match, < 25% id, < 50% id, < 75% id and > 75% id) divided by the total length of the reference sequence. As the authors state, this "*gives a realistic view of the overall reference and query similarity which is often not very precisely measured through visual inspection*" (Cabanettes & Klopp, 2018).

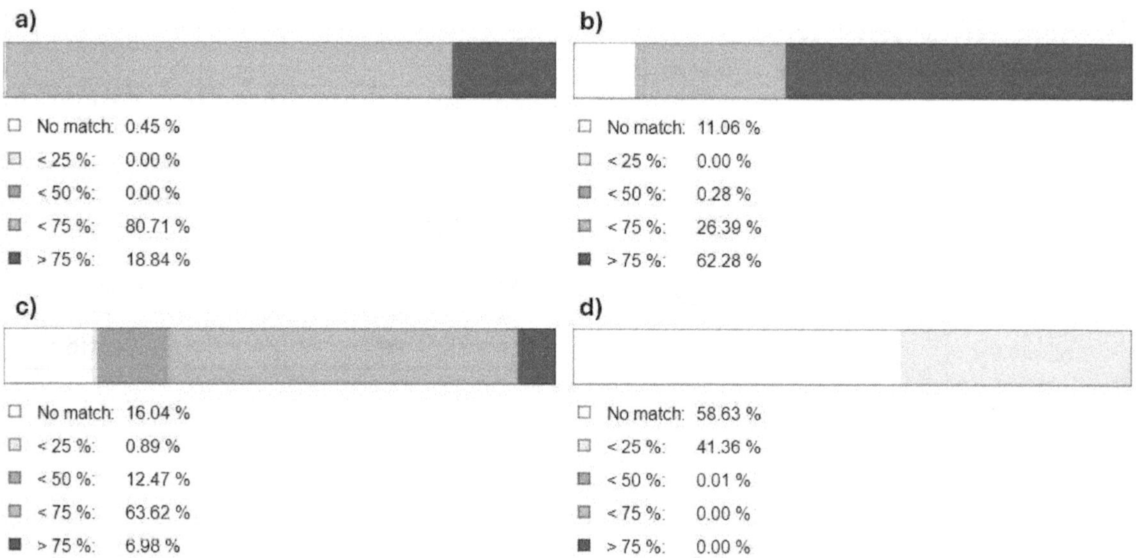

Figure 52. Identity summaries for the dot plot comparisons shown in previous figures: a) *Haemophilus influenzae* strains PittEE and PittGG (**Figure 34**); b) chromosome 1 of *Homo sapiens* and *Pan troglodytes* (**Figure 34**), *c)* Asian wild rice *Oryza rufipogon* and the cultivated *Oryza sativa* subspecies Japonica (**Figure 47**); d) *Aphanius iberus* and *Poecilia reticulata* (**Figure 49**).

The identity summary bar plot is generated by clicking on the button marked "Summary" on the D-GENIES results page. When doing so for four of the plots that we have discussed previously in this text, we got the results shown in **Figure 52**. We can see that the two *H. influenzae* strains had highly similar genome sequences, with more than 99% of the sequence being in the ranges of 50-75% identical (81% of the sequence) or >75% identical (19% of the sequence). In the comparison of human and chimpanzee chromosome 1, about 89% of the sequence were in the two upper ranges of identity, with 26% being 50-75% identical and 62% being >75% identical. Further, the two rice genomes had slightly lower overall similarity, with 70% of the sequence being in the ranges 50-75% identical (64%) or >75% identical (7%). However, and quite surprisingly, given the convincingly looking dot plot in **Figure 49**, overall sequence similarity between *A. iberus* and *P. reticulata* turned out to be drastically lower. Distinctly more than half of the sequence, 59%, had no match, and almost all of the remainder of the sequence was <25% identical. To understand this seemingly surprising statistic, we need to look more closely at the *A. iberus* versus *P. reticulata* dot plot.

In **Figure 53** (left panel) we have marked the ends of *P. reticulata* chromosome 12 with vertical lines in the overall dot plot. This view gives the impression that there is a continuous match covering almost the full length of chromosome 12, appearing as an unbroken green line, also representing a high level of sequence identity (the settings were adjusted to only display matches with at least 75% identity). However, when clicking on this region to zoom in, we see the plot in the right panel of **Figure 53**. It now becomes evident that the seemingly contiguous line actually consists of many short matches, so that the diagonal is barely visible. In the overall plot, representing much longer sequences, single pixels representing short high-identity matches will float together to give the appearance of an unbroken diagonal, giving a misleading impression of the sequence identity being higher than what it actually is. This example clearly illustrates the point made by the authors of D-GENIES, when they stated that sequence similarity "*is often not very precisely measured through visual inspection*" and that the statistics presented by the identity summary tool should be used for this purpose.

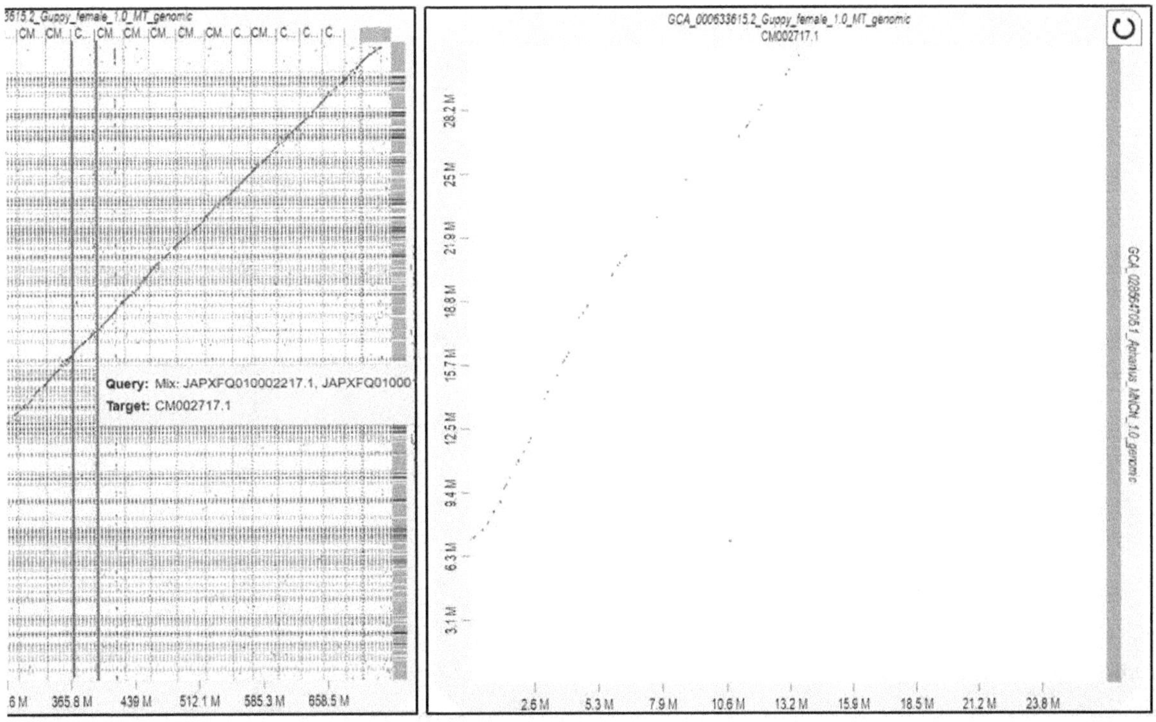

Figure 53. Left panel: Partial view of the overall dot plot between *Aphanius iberus* and *Poecilia reticulata* with the borders of *P. reticulata* chromosome 12 marked with vertical lines. Approximately a third of the plot was cut away to make the information in the figure of readable size. Right panel: Zoom on the dotplot for *P. reticulata* chromosome 12 versus the best matching *A. iberus* scaffold.

In addition to using the identity summary and manually inspecting parts of the plot by zooming in to chromosome, scaffold or contig level, is to use filtering or noise removal. The D-GENIES results page has two functions for these purposes. The user can change the "Filter small matches" to different levels to remove fewer or more of the short matches. Using the "Remove noise" button instantly removes every match that is considered as noise according to the definition used by the algorithm. Briefly described, a match will be considered as "noise" if it is short and if matches of the same length occur very frequently in the data set. In our example, if we apply the "Remove noise" function to the plot that was shown in **Figure 49**, all matches will disappear, which again demonstrates that what looks like a solid green diagonal across the whole plot is actually a series of very short matches with gaps in between.

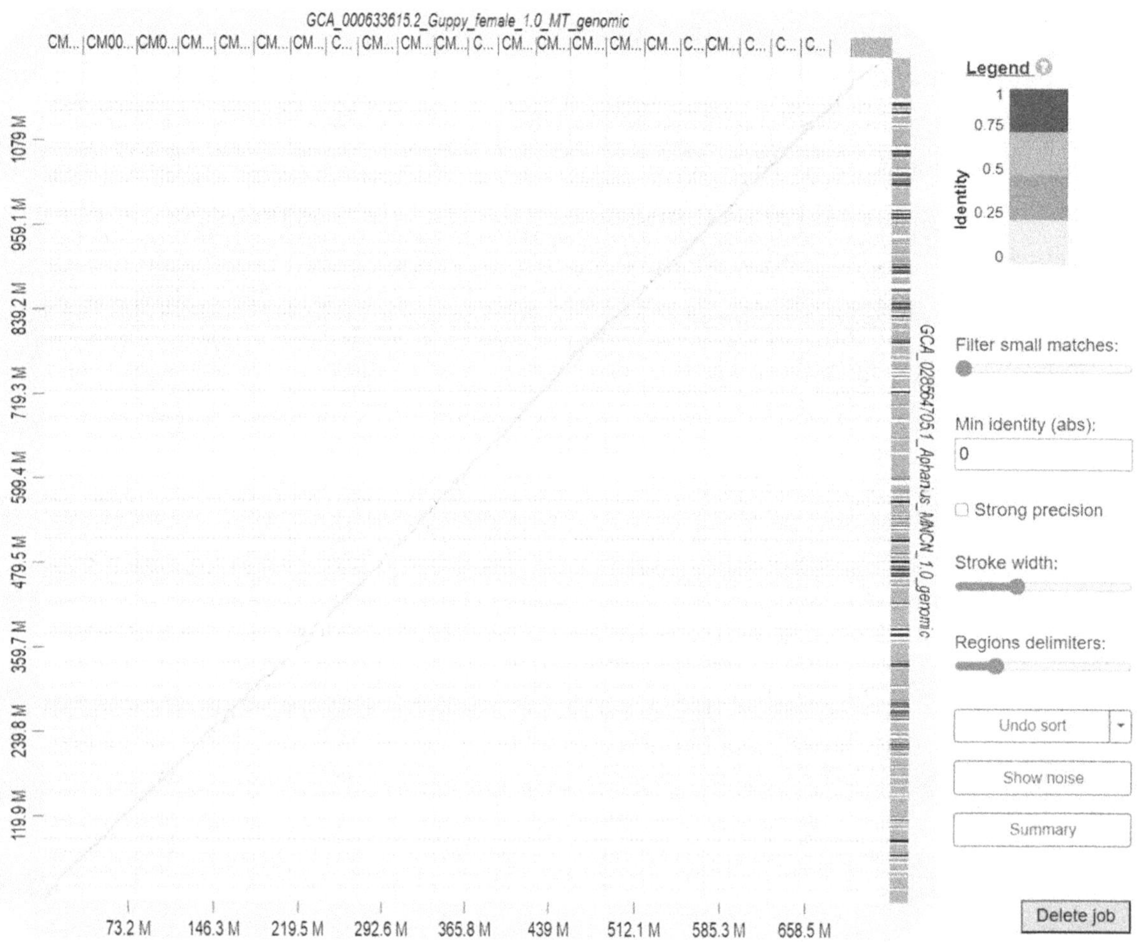

Figure 54. Dot plot between *Aphanius iberus* and *Poecilia reticulata* with noise removed and minimum identity adjusted to 0%.

However, when lowering the "Min identity" setting, the diagonal reappears, but is now shown in yellow in **Figure 54**. This demonstrates that almost all of the significant matches have low sequence identity, below 25%. Now comparing the adjusted plot (**Figure 54**) with the identity summary (**Figure 52 d**) one could argue that the visual impression is still somewhat misleading, since the summary shows that about 58% of the sequence has no match. This once again illustrates that it is better to trust the statistics than the visual impression when estimating sequence similarity.

As we have seen in the various examples described so far, dot plots are highly useful for identifying and visualising structural properties, such as repeat regions, gene duplications, inversions, translocations, deletions and inserts. However, there are also risks with relying on visual interpretation of sequence properties, in particular overall sequence similarity, as the last example illustrates.

A brief history of research on dot plot methods

The first scientific article that presented a dot plot method for pairwise comparison of biological sequences seems to be the 1970 paper by Gibbs and McIntyre, titled "The Diagram, a Method for Comparing Sequences" (Gibbs & Mcintyre, 1970). The authors cited an even earlier article (Fitch, 1969) as the inspiration for their method, but it seems that the "Diagram" method proposed by Gibbs and McIntyre is the first example of a dot plot sequence comparison method. The proposed method was to record the two sequences along adjacent sides of a rectangular matrix, which they called "the diagram", and to mark with a dot every location where the row and column contain matching amino acids or nucleotides. Using, as their initial example, cytochrome c protein sequences from human, rhesus monkey, tuna fish and a bacterium, the authors showed how the presence or scarcity of diagonal rows of matches indicates whether the sequences are closely or distantly related (Gibbs & Mcintyre, 1970, Figure 1).

Interestingly, these authors also proposed methods for determining if the number of diagonal matches is significantly higher than expected by chance, so that the dot plot, or "diagram", would output a numerical measure of sequence similarity in addition to the visual comparison. In addition, the article suggests that the comparisons of sequences "*can be done by hand, but are done more quickly by computer*" (Gibbs & Mcintyre, 1970). They also provided access to a computer program, named MATCH, implemented in Fortran, for generating the "diagram" and computing the similarity measures.

When the idea of plotting sequences in a two-dimensional display had become an established technique for sequence comparison, researchers began working on enhancements of the visualisation. An early example of this is the "graphic matrix procedure" proposed by (Maizel & Lenk, 1981). Their program, named REVEAL, was implemented in BASIC for a desktop computer with 64 kilobytes of memory and an integral dot matrix printer, i.e. the type of printer commonly used with the first personal computers. It made use of four colours when displaying plots of nucleotide sequences and a "filtration procedure" to allow a dot in the matrix to represent several positions, which is essentially the windows-based matching technique. The authors summed up the advantages of dot plots by stating that the "*graphic matrix technique takes advantage of the sensitive discrimination of the eye and the mind in detecting patterns in nucleic acid and protein sequences*" (Maizel & Lenk, 1981) and demonstrated its use to detect features such as RNA secondary structure and protein coding regions. They also showed that the "filtration" technique made it possible to display a dot plot at the scale of small genomes, using the approximately 5000 bases long genomes of simian virus 40 and polyoma virus as examples.

In the compressed graphic matrix for these two sequences, comprising a total of 5336 x 5160 = 26,966,160 nucleotide comparisons, each dot indicated that five nucleotides in a row (window) were identical (Maizel & Lenk, 1981, Figure 4).

In 1982 the computer program DIAGON was presented in an article published in the journal Nucleic Acids Research (Staden, 1982). This program was based on the Gibbs and McIntyre method and also implemented in Fortran, but offered more elaborate methods for similarity scoring and an interactive interface through which the user could make various manipulations. It was, for example, possible to make insertions in sequences so that alignments were obtained, which could be downloaded and saved to disk. The author stated in the article that readers could request a "*magnetic tape containing DIAGON*" for usage in their laboratories to analyse their sequences (Staden, 1982). A more efficient implementation of DIAGON was later developed in Pascal, which could be run on microcomputers, and boasted a runtime of 15 minutes for comparison of protein sequences of length 500 amino acids (Giles, 1984).

Through the 1980s several articles and computer program implementations addressed the issues of how to distinguish significant matches from random ones and how to visualise the biologically relevant similarities. For example, Reisner and Buchholtz experimented with various methods for scoring amino acid similarities and colour schemes for visualising significant matches (Reisner & Bucholtz, 1988). Their program, implemented in Fortran, used colour as a "*quasi-third dimension*" in the two-dimensional dot plots. This was suggested as a remedy for the limitation that when "*using a scoring matrix the appearance of a dot in the output gives no indication per se as to score value for that particular match – only that it fell above the threshold*" (Reisner & Bucholtz, 1988).

In the subsequent years, several attempts were made to speed up the processing so that longer sequences could be analysed in reasonable time. One prominent example is the word-based method published by Lefévre & Lkeda (1994). They proposed to use a data structure named position trees to represent the positions in the sequence for every word. After pre-compiling such a tree for each sequence, matches were identified by a tree traversal algorithm, 'Match2Trees'. The authors demonstrated that this algorithm made it possible to scale up the dot plot technique to genomic sequences. The article showcased an example using the full sequence of *S. cerevisiae* chromosome 3, with a length of 315,357 nucleotides (ENA accession number X59720), demonstrating that their program could generate a dot plot of this sequence against itself in 12 minutes. Although considered fast for that time, this was still very slow in comparison to the later programs Gepard and D-GENIES, which both plot the X59730 sequence against itself in less than a second.

When presenting the DOTTER program, Sonnhammer & Durbin (1995) noted that previous programs using words or windows had the limitation that the similarity threshold for drawing a dot representing a match had to be set *a priori*. When an unsuitable threshold was chosen, the plot would either be "*full of noise or lacking the relevant diagonals*" (Sonnhammer & Durbin, 1995). Upon observing a plot having one of these problems, the user would need to either increase or lower the threshold and then re-run the analysis, which usually required a time-consuming recalculation of the whole matrix. The main novelty in DOTTER was that the similarity threshold could be adjusted by the user after the calculation of the dot matrix.

The DOTTER tool uses two thresholds that can be set dynamically by the user to determine how dots are drawn using a 128-colour greyscale. The first (lower) threshold determines if a dot is drawn or not, whereas the second (higher) threshold sets the score at which the dot is drawn with maximum intensity. For all scores in the interval between the lower and higher threshold, a dot is drawn with an intensity proportional to the score for the window it represents. DOTTER was implemented in C and made available for installation in UNIX environments.

In 2004 Java-Dotter (JDotter[26]), a platform-independent Java interface for DOTTER, was presented (Brodie et al., 2004). JDotter works as a client-server application, which sends sequences to the Linux version of DOTTER, which generates the dot matrix on a central server, then sends the plot back to the client for interactive viewing in the local Java application. In **Figure 55** we show a JDotter dot plot of the human and rat S100A3 protein sequences, which we previously plotted with Dotmatcher in **Figure 12** and with Gepard in **Figure 27**. The same clear diagonal running across approximately half of the sequence length is seen in the JDotter plot when shown with default settings. In the Jdotter plot in **Figure 55**, the GreyMap Tool settings were changed by adjusting the dot intensity to the range 1 to 70. This made the main diagonal clearer and also made the additional, shorter diagonal starting at approximately the coordinates (440, 420) visible, which was not seen before the intensity range adjustment.

[26] JDotter was developed by researchers at the University of Victoria, Canada, and was at the time of writing available for download from the following web page:
4virology.net/virology-ca-tools/jdotter/

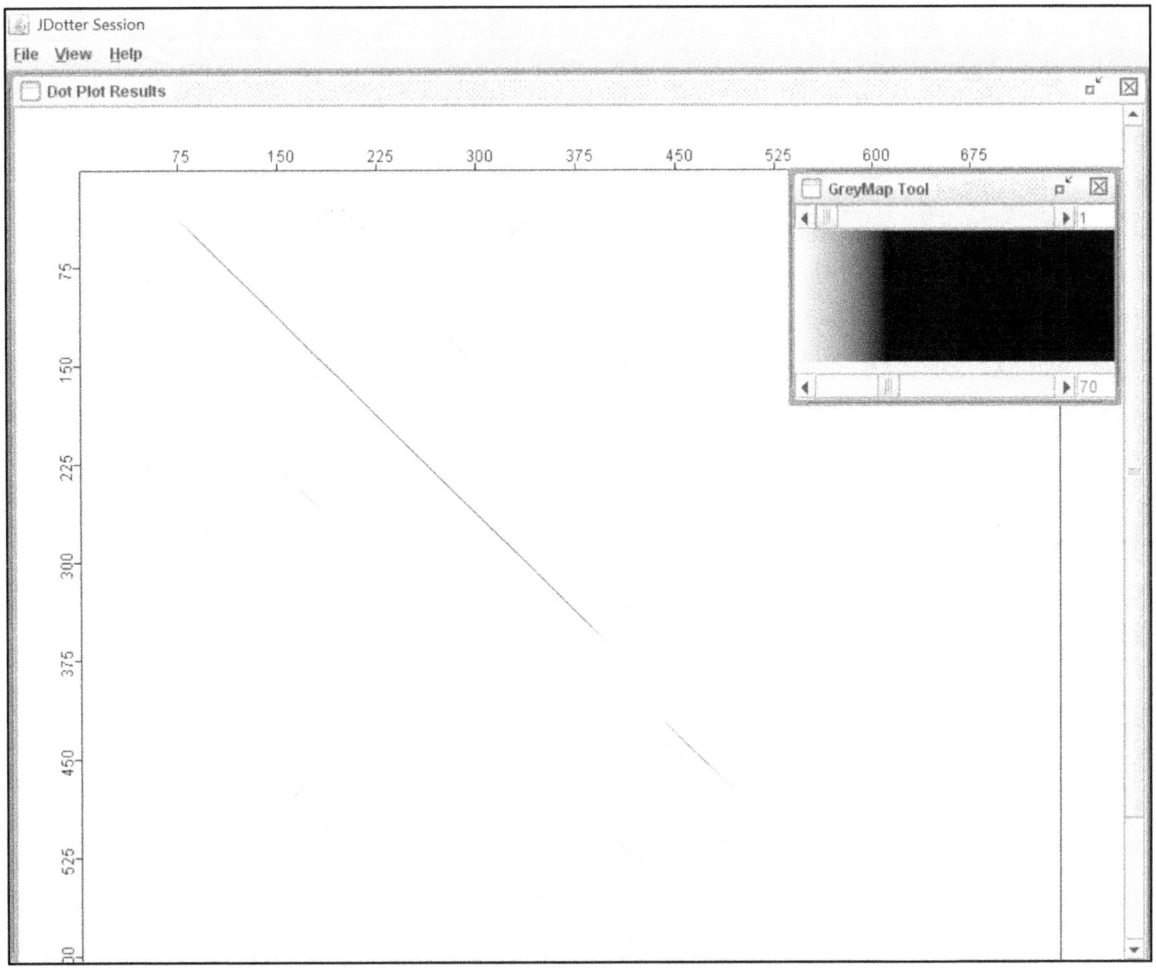

Figure 55. Dot plot of mRNA sequences from the human (vertical axis) and rat (horizontal axis) S100A3 gene, encoding the S100 calcium-binding A3 protein. Generated with JDotter, using default parameter settings. The GreyMap Tool was used to make the main diagonal clearer by adjusting the intensity range setting to 1-70.

Overall, the interactive interface of JDotter has similar capabilities for adjusting the thresholds for visualisation as those in Gepard. In general, JDotter and Gepard offer the same advantages of ease of use and platform independence and are therefore both good choices for making dot plot comparisons of genes or protein sequences. The main difference between the two is in terms of speed of processing when plotting longer sequences. The algorithm for generating dot matrices in DOTTER and JDotter uses the windows-based approach, which we introduced previously and exemplified with the Dotmatcher program. In the original article on DOTTER, the authors describe how their algorithm calculates the summed score for each window *"in a sliding manner by simply adding the next score and subtracting the last score inside the window"*. To speed up these calculations of window score sums, they used precalculated score vectors for every possible symbol in the vertical sequence along the horizontal sequence – an idea adapted from an earlier published dot plot program (Karreman, 1992). Also, instead of calculating the window sums for one diagonal at a time, the DOTTER algorithm keeps a horizontal vector of all window sums, to which the precalculated score vectors are added and subtracted row by row.

Despite these attempts to speed up the calculations, DOTTER is substantially slower than modern software, for example Gepard and D-GENIES, since it does not make use of advanced algorithmic concepts such as suffix matrices or minimizers. The difference in speed between Gepard and JDotter is not noticeable for short sequences, but becomes significant when genome-size sequences are compared. For example, the Mycoplasma example plot in **Figure 28** was generated by Gepard in a few seconds, despite the sequences being about 0.5 Mbp long. When generating the corresponding plot with JDotter, we noted a runtime of approximately 20 minutes. The resulting plot from JDotter in **Figure 56** was made with the GreyMap range set to 0-100, resulting in a plot similar to the leftmost one from Gepard in **Figure 30**. In other words, the main difference between the two tools when run on this example data is Gepard's approximately 100-fold increase in speed when run with the suffix array option.

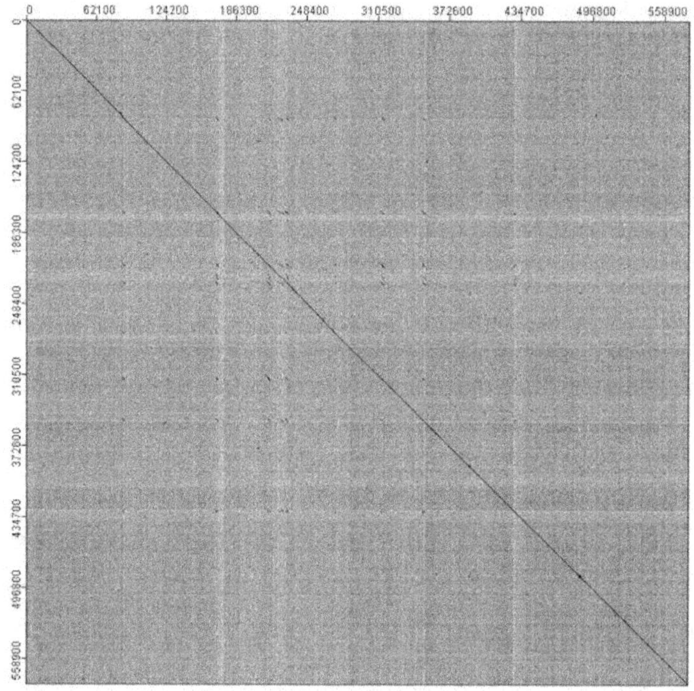

Figure 56. JDotter dot plot comparison of the full genome sequences of two Mycoplasma genitalium strains, for comparison with the Gepard results in **Figure 28** and **Figure 30**.

Among the attempts to speed up dot plot construction, the method published by Huang & Zhang (2004) should be mentioned. These authors cited the position tree method (Lefévre & Ikeda, 1994) and aimed to develop a similar but faster method. They suggested to build a k-tuple lookup table for each sequence, showing the position of every word in the sequence – a representation similar to a hash table. They noted that if the alphabet consists of p

elements (e.g. for nucleotide sequences, $p = 4$) and the word length is set to k, there are p^k possible different k-tuples. If each k-tuple is represented by an integer from 1 to p^k, a one-dimensional array C of length p^k can be used to store the last position of each k-tuple in the sequence. This is illustrated in **Figure 57**, where we have an example sequence of 15 nucleotides, the 16 k-tuples AA, AC, ..., TT, for $k = 2$, and the array C storing the last position of each k-tuple in the sequence. The background colouring of cells should help the interpretation. Lighter colour marks the k-tuple AC, with code = 2. Thus, $C[2] = 6$ since the last (and only) occurrence of AC is in position six in the sequence. Darker colour marks the k-tuple AG, with code = 3. This k-tuple occurs twice in the sequence at positions 8 and 14. In the C array we have $C[3] = 14$ since the last occurrence of AG is in position 14 and the C array only records the last occurrence. In cases where a k-tuple does not occur anywhere in the sequence, the corresponding value in C will be 0. From this example, it should be clear that when we observe a particular k-tuple in one sequence, we can find the last occurrence of the same k-tuple in the other sequence by simply checking the value in the corresponding cell in C.

Sequence	A	T	G	A	C	A	C	A	G	C	T	T	C	A	G	
Position	1	2	3	4	5	6	7	8	9	10	11	12	13	14	15	
k-tuples	AA	AC	AG	AT	CA	CC	CG	CT	GA	GC	GG	GT	TA	TC	TG	TT
Codes	1	2	3	4	5	6	7	8	9	10	11	12	13	14	15	16
Initial D	4	15	9	2	5	2	5	3	10	8	16	14	5	3		
Final D	0	0	0	0	0	4	5	0	0	0	0	0	7	8		
Initial C	0	0	0	0	0	0	0	0	0	0	0	0	0	0	0	0
Final C	0	6	14	1	13	0	0	8	3	9	0	0	0	12	2	11

Figure 57. Lookup table to find the last occurrence of every possible k-tuple in a sequence. Example sequence consisting of 15 nucleotides, with position numbers 1-15. For $k = 2$, there are a total of 16 possible k-tuples, AA, ..., TT, with codes 1-16. The one-dimensional array C is initially filled with zeroes. When table construction (which is explained later) has been finalized, each element of the C array will contain a number representing the last occurrence in the sequence of the corresponding k-tuple. Light background colour: The last occurrence of AC in the sequence is at position 6, therefore $C[2] = 6$. Darker background colour: The last occurrence of AG is at position 14, therefore $C[3] = 14$.

In addition to the array C, there is a similar one-dimensional array D of length L-k+1 (where L is the length of the sequence). Each element $D[i]$ in this array initially just stores the coded form, $c(i)$, of the k-tuple occurring at position i in the sequence. The initial value of $D[1]$ is therefore 4 since AT, the k-tuple at position 1, has the code 4. The initial value of $D[2]$ = 15, since the k-tuple at position 2, TG, has code 15, and so on. In the array C, all values are initially set to 0. Given these initial values in the C and D arrays, the lookup table can be generated in a single pass through the sequence. Each time a k-tuple is found in a new location, position j, the previous position $C[c(i)]$ will be assigned to D[j] and j will be assigned to $C[c(i)]$. In practice, this works as follows. When we start at the beginning of the sequence and find AT at position 1, the value of $D[1]$, which was initially 4, will be set to 0 and $C[4]$ will be set to 1. Next, when TG is found at position 2 in the sequence, the value of $D[2]$ will change from 15 to 0 and $C[15]$ will change from 0 to 2. Looking at these two k-tuples as examples, we can now see how the final C and D arrays show the locations of AT and TG in the sequence. The value $C[4] = 1$ represents that the last occurrence of AT is at position 1 and the value $D[1] = 0$ represents that there are no other occurrences of AT. Similarly, the value $C[15] = 2$ shows that the last occurrence of TG is at position 2 and $D[2] = 0$ represents that this is the only TG di-nucleotide in the sequence.

Looking a bit further ahead in the process of generating the lookup table, we move on to position five in the sequence, where the k-tuple CA is found. The value of $D[5]$ will then be set to 0 (the "preceding position" of CA), and $C[5]$ will be set to 5 (representing the occurrence of CA at position 5). When the second occurrence of CA is found at position 7, $D[7]$ will be set to 5 (the preceding position of CA) and $C[5]$ will be updated to 7. Finally, when the third occurrence of CA is found at position 13, $D[13]$ will be set to 7 and $C[5]$ will be updated to 13. In this way, the final version of the two arrays C and D will essentially form a chain of pointers showing us the three locations of CA.

When using the lookup table, which consists only of the final versions of the two arrays C and D, to find all occurrences of CA, we start in $C[5]$, which shows that the last occurrence is in position 13. Then looking in $D[13]$, we see that the next-to-last occurrence is in position 7, and looking in $D[7]$ we find that the occurrence before that one is in position 5. Finally, when looking in $D[5]$ we find the value 0, indicating that there are no more occurrences of CA in the sequence. This procedure is illustrated in **Figure 58**.

Sequence	A	T	G	A	**C**	**A**	**C**	**A**	G	C	T	T	**C**	**A**	G	
Position	1	2	3	4	**5**	6	**7**	8	9	10	11	12	**13**	14	15	
k-tuples	AA	AC	AG	AT	CA	CC	CG	CT	GA	GC	GG	GT	TA	TC	TG	TT
Codes	1	2	3	4	5	6	7	8	9	10	11	12	13	14	15	16
D	0	0	0	0	**0**	4	**5**	0	0	0	0	0	**7**	8		
C	0	6	14	1	**13**	0	0	8	3	9	0	0	0	12	2	11

Figure 58. Illustrating that the final versions of the two arrays C and D contain sufficient information to trace all occurrences of a given k-tuple. In the example, we locate all occurrences of CA, which has the code 5. We therefore start in $C[5]$, which points to position 13 in the sequence where the last occurrence of CA is located. Then observing that $D[13] = 7$, we find that position 7 in the sequence also contains an occurrence of CA, then $D[7] = 5$ leads us to find CA also at position 5. Given that $D[5] = 0$, we conclude that there are no more occurrences of this k-tuple and stop the search. This procedure obviously requires substantially fewer steps than scanning through the whole sequence to find all occurrences of CA.

When a lookup table has been generated for each of the two sequences, as described above and in the article (Huang & Zhang, 2004), it is straightforward to find all k-tuple matches between the two compared sequences. However, the lookup tables grow exponentially for high values of k, so that a relatively low k is necessary to not use up too much memory. Also, for high values of k, the method would lack sensitivity to highly similar but non-identical matches. The authors therefore developed a two-step matching process, where the initial step is to identify all k-tuple matches. In the second step, for each k-tuple match, the sequences are compared in a window of length w, where $w > k$. In this step mismatches were allowed, with a scoring scheme giving 0 for identical nucleotides and -1 for mismatches, and a negative score threshold s, in essence allowing up to s number of mismatches for the w positions in the window. In the examples shown in the article, the authors used k-tuple sizes from 5 to 12, window lengths from 27 to 69, and allowed 0 to 4 mismatches (Huang & Zhang, 2004).

The method described above is reasonably fast and comparable to the suffix array-based method used in Gepard. The authors state that the most time-consuming of the three steps of the algorithm (i.e. lookup table generation, k-tuple search, and window-based evaluation of matches) was the k-tuple search phase (Huang & Zhang, 2004). Runtimes for the examples shown in the article seem similar to those for Gepard. For example, the authors plotted the two genomes of *Thermoplasma volcanium* and *Thermoplasma acidophilum* (Genbank accession numbers BA000011 and AL139299) which are each approximately 1.5 Mb long. Using $k = 11$, $w = 27$ and $s = -4$, the runtime was only 4.8 seconds on a personal computer with a 1 GHz processor and 640 Mb memory. When reproducing the plot using Gepard on a modern laptop computer, we noted a similar runtime. However, the method by Huang & Zhang (2004) does not scale to larger genomes as well as D-GENIES does. The article presented a case where a 13 Mbp long contig of mouse chromosome 9 was plotted against a 24 Mbp contig of human chromosome 11. Runtime for this example was reported as 8 minutes. When reproducing this plot using the online version of D-GENIES, we found a runtime of only 4 seconds, which again illustrates the immense speedup gained by the usage of minimizers.

The software implementation of the method developed by Huang & Zhang (2004) does not seem to be freely available currently. In the article, the authors referred to the website www.lynnon.com for access to the program. At the time of writing, this site provides a commercial bioinformatics software package named DNAMAN which, among many other modules, includes a dot plot tool[27]. This software, i.e. DNAMAN, has been cited in various research articles presenting dot plots, e.g. Magee et al. (2010). Other authors who presented dot plots and cited Huang & Zhang (2004) have referred to the software as LBDot (Huang et al., 2017; Nakaya et al., 2017; Wan et al., 2009) or LBDotView (Craveiro et al., 2015).

[27] www.lynnon.com/homology.html#dot

Bibliography

Abdullah, M., Furtado, A., Masouleh, A. K., Okemo, P., & Henry, R. (2025). The genomes of the most diverse AA genome rice species provide a resource for rice improvement and studies of rice evolution and domestication. *BMC Genomics*, *26*(1), 54. https://doi.org/10.1186/S12864-025-11246-0

Altschul, S. F., & Erickson, B. W. (1986). Optimal sequence alignment using affine gap costs. *Bulletin of Mathematical Biology*, *48*(5–6), 603–616. https://doi.org/10.1007/BF02462326

Altschul, S. F., Gish, W., Miller, W., Myers, E. W., & Lipman, D. J. (1990). Basic local alignment search tool. *Journal of Molecular Biology*, *215*(3), 403–410. https://doi.org/10.1016/S0022-2836(05)80360-2

Arita, M., Karsch-Mizrachi, I., & Cochrane, G. (2021). The international nucleotide sequence database collaboration. *Nucleic Acids Research*, *49*(D1), D121–D124. https://doi.org/10.1093/NAR/GKAA967

Armstrong, J., Fiddes, I. T., Diekhans, M., & Paten, B. (2018). Whole-Genome Alignment and Comparative Annotation. *Annual Review of Animal Biosciences*, *7*, 41. https://doi.org/10.1146/ANNUREV-ANIMAL-020518-115005

Azam, S., Sahu, A., Kadivella, M., Khan, A. W., Neupane, M., Tassell, C. P. Van, Rosen, B. D., Gandham, R. K., Rath, S. N., & Majumdar, S. S. (2025). Genome assemblies of Indian desi cattle reveals hotspots of rearrangements and immune-related genetic diversity. *BioRxiv*, 2025.01.14.632934. https://doi.org/10.1101/2025.01.14.632934

Benson, D. A., Cavanaugh, M., Clark, K., Karsch-Mizrachi, I., Lipman, D. J., Ostell, J., & Sayers, E. W. (2013). GenBank. *Nucleic Acids Research*, *41*(Database issue). https://doi.org/10.1093/NAR/GKS1195

Blum, M., Andreeva, A., Cavalcanti Florentino, L., Rocio Chuguransky, S., Grego, T., Hobbs, E., Lazaro Pinto, B., Orr, A., Paysan-Lafosse, T., Ponamareva, I., Salazar, G. A., Bordin, N., Bork, P., Bridge, A., Colwell, L., Gough, J., Haft, D. H., Letunic, I., Llinares-López, F., ... Bateman, A. (2025). InterPro: the protein sequence classification resource in 2025. *Nucleic Acids Research*, *53*(D1), D444–D456. https://doi.org/10.1093/NAR/GKAE1082

Brady, T. S., Merrill, B. D., Hilton, J. A., Payne, A. M., Stephenson, M. B., & Hope, S. (2017). Bacteriophages as an alternative to conventional antibiotic use for the prevention or treatment of Paenibacillus larvae in honeybee hives. *Journal of Invertebrate Pathology, 150,* 94–100. https://doi.org/10.1016/J.JIP.2017.09.010

Brodie, R., Roper, R. L., & Upton, C. (2004). JDotter: a Java interface to multiple dotplots generated by dotter. *Bioinformatics, 20*(2), 279–281. https://doi.org/10.1093/BIOINFORMATICS/BTG406

Cabanettes, F., & Klopp, C. (2018). D-GENIES: Dot plot large genomes in an interactive, efficient and simple way. *PeerJ, 2018*(6), e4958. https://doi.org/10.7717/PEERJ.4958

Castro, C. J., & Ng, T. F. F. (2017). U50: A New Metric for Measuring Assembly Output Based on Non-Overlapping, Target-Specific Contigs. *Journal of Computational Biology, 24*(11), 1071. https://doi.org/10.1089/CMB.2017.0013

Craveiro, S. R., Inglis, P. W., Togawa, R. C., Grynberg, P., Melo, F. L., Ribeiro, Z. M. A., Ribeiro, B. M., Báo, S. N., & Castro, M. E. B. (2015). The genome sequence of Pseudoplusia includens single nucleopolyhedrovirus and an analysis of p26 gene evolution in the baculoviruses. *BMC Genomics, 16*(1), 1–13. https://doi.org/10.1186/S12864-015-1323-9

Das, A., & Schatz, M. C. (2022). Sketching and sampling approaches for fast and accurate long read classification. *BMC Bioinformatics, 23*(1), 1–23. https://doi.org/10.1186/S12859-022-05014-0

de Castro, E., Sigrist, C. J. A., Gattiker, A., Bulliard, V., Langendijk-Genevaux, P. S., Gasteiger, E., Bairoch, A., & Hulo, N. (2006). ScanProsite: detection of PROSITE signature matches and ProRule-associated functional and structural residues in proteins. *Nucleic Acids Research, 34*(Web Server issue). https://doi.org/10.1093/NAR/GKL124

Dong, M. J., Luo, H., & Gao, F. (2022). Ori-Finder 2022: A Comprehensive Web Server for Prediction and Analysis of Bacterial Replication Origins. *Genomics, Proteomics & Bioinformatics, 20*(6), 1207–1213. https://doi.org/10.1016/J.GPB.2022.10.002

Doolittle, R. F. (1995). The multiplicity of domains in proteins. *Annual Review of Biochemistry, 64*, 287–314. https://doi.org/10.1146/ANNUREV.BI.64.070195.001443

Ebersberger, I., Metzler, D., Schwarz, C., & Pääbo, S. (2002). Genomewide comparison of DNA sequences between humans and chimpanzees. *American Journal of Human Genetics, 70*(6). https://doi.org/10.1086/340787

Eddy, S. R. (2004). What is dynamic programming? *Nature Biotechnology 2004 22:7, 22*(7), 909–910. https://doi.org/10.1038/nbt0704-909

Ekundayo, B., & Bleichert, F. (2019). Origins of DNA replication. *PLoS Genetics*, *15*(9), e1008320. https://doi.org/10.1371/JOURNAL.PGEN.1008320

Fass, D. (2012). Disulfide bonding in protein biophysics. *Annual Review of Biophysics*, *41*(1), 63–79. https://doi.org/10.1146/ANNUREV-BIOPHYS-050511-102321

Fitch, W. M. (1969). Locating gaps in amino acid sequences to optimize the homology between two proteins. *Biochemical Genetics*, *3*(2), 99–108. https://doi.org/10.1007/BF00520346

Fleischmann, R. D., Adams, M. D., White, O., Clayton, R. A., Kirkness, E. F., Kerlavage, A. R., Bult, C. J., Tomb, J. F., Dougherty, B. A., Merrick, J. M., McKenney, K., Sutton, G., FitzHugh, W., Fields, C., Gocayne, J. D., Scott, J., Shirley, R., Liu, L. I., Glodek, A., … Venter, J. C. (1995). Whole-genome random sequencing and assembly of Haemophilus influenzae Rd. *Science*, *269*(5223), 496–512. https://doi.org/10.1126/SCIENCE.7542800

Gasteiger, E., Hoogland, C., Gattiker, A., Duvaud, S., Wilkins, M. R., Appel, R. D., & Bairoch, A. (2005). Protein Identification and Analysis Tools on the ExPASy Server. *The Proteomics Protocols Handbook*, 571–607. https://doi.org/10.1385/1-59259-890-0:571

Gibbs, A. J., & Mcintyre, G. A. (1970). The Diagram, a Method for Comparing Sequences. *European Journal of Biochemistry*, *16*(1), 1–11. https://doi.org/10.1111/J.1432-1033.1970.TB01046.X

Giles, I. G. (1984). A microcomputer implementation of the DIAGON sequence comparison and alignment program. *Biochemical Society Transactions*, *12*(6), 1101–1101. https://doi.org/10.1042/BST0121101

Hanson, R. M., Prilusky, J., Renjian, Z., Nakane, T., & Sussman, J. L. (2013). JSmol and the Next-Generation Web-Based Representation of 3D Molecular Structure as Applied to Proteopedia. *Israel Journal of Chemistry*, *53*(3–4), 207–216. https://doi.org/10.1002/IJCH.201300024

Hardison, R. C. (2012). Evolution of Hemoglobin and Its Genes. *Cold Spring Harbor Perspectives in Medicine*, *2*(12), a011627. https://doi.org/10.1101/CSHPERSPECT.A011627

Henikoff, S., & Henikoff, J. G. (1992). Amino acid substitution matrices from protein blocks. *Proceedings of the National Academy of Sciences of the United States of America*, *89*(22), 10915–10919. https://doi.org/10.1073/PNAS.89.22.10915

Hogg, J. S., Hu, F. Z., Janto, B., Boissy, R., Hayes, J., Keefe, R., Post, J. C., & Ehrlich, G. D. (2007). Characterization and modeling of the Haemophilus influenzae core and supragenomes based on the complete genomic sequences of Rd and 12 clinical nontypeable strains. *Genome Biology*, *8*(6). https://doi.org/10.1186/GB-2007-8-6-R103

Huang, Y., Yu, F., Li, X., Luo, L., Wu, J., Yang, Y., Deng, Z., Chen, R., & Zhang, M. (2017). Comparative genetic analysis of the 45S rDNA intergenic spacers from three Saccharum species. *PLOS ONE*, *12*(8), e0183447. https://doi.org/10.1371/JOURNAL.PONE.0183447

Huang, Y., & Zhang, L. (2004). Rapid and sensitive dot-matrix methods for genome analysis. *Bioinformatics (Oxford, England)*, *20*(4), 460–466. https://doi.org/10.1093/BIOINFORMATICS/BTG429

IUPAC Blue Book. (2022). *Chemistry International*, *44*(3). https://doi.org/10.1515/ci-2022-0314

Jumper, J., Evans, R., Pritzel, A., Green, T., Figurnov, M., Ronneberger, O., Tunyasuvunakool, K., Bates, R., Žídek, A., Potapenko, A., Bridgland, A., Meyer, C., Kohl, S. A. A., Ballard, A. J., Cowie, A., Romera-Paredes, B., Nikolov, S., Jain, R., Adler, J., ... Hassabis, D. (2021). Highly accurate protein structure prediction with AlphaFold. *Nature 2021 596:7873*, *596*(7873), 583–589. https://doi.org/10.1038/s41586-021-03819-2

Karreman, C. (1992). A dotplot program for the Atari ST, for the analysis of DNA and protein sequences. *Computer Applications in the Biosciences : CABIOS*, *8*(1), 75–77. https://doi.org/10.1093/BIOINFORMATICS/8.1.75

Khemaissa, S., Sagan, S., & Walrant, A. (2021). Tryptophan, an Amino-Acid Endowed with Unique Properties and Its Many Roles in Membrane Proteins. *Crystals 2021, Vol. 11, Page 1032*, *11*(9), 1032. https://doi.org/10.3390/CRYST11091032

Kok, D. N., Gosselin, S. P., Howard, B., Cresawn, S. G., Tsourkas, P. K., & Hendrickson, H. L. (2025). Genomic Analysis of 96 Paenibacillus larvae Bacteriophages Including 26 from Aotearoa, New Zealand. *Viruses*, *17*(2). https://doi.org/10.3390/V17020137

Krumsiek, J., Arnold, R., & Rattei, T. (2007). Gepard: a rapid and sensitive tool for creating dotplots on genome scale. *Bioinformatics*, *23*(8), 1026–1028. https://doi.org/10.1093/BIOINFORMATICS/BTM039

Kumar, S., Suleski, M., Craig, J. M., Kasprowicz, A. E., Sanderford, M., Li, M., Stecher, G., & Hedges, S. B. (2022). TimeTree 5: An Expanded Resource for Species Divergence

Times. *Molecular Biology and Evolution*, *39*(8). https://doi.org/10.1093/MOLBEV/MSAC174

Kurochkina, N., & Guha, U. (2012). SH3 domains: modules of protein–protein interactions. *Biophysical Reviews*, *5*(1), 29. https://doi.org/10.1007/S12551-012-0081-Z

Lefévre, C., & Ikeda, J. E. (1994). A fast word search algorithm for the representation of sequence similarity in genomic DNA. *Nucleic Acids Research*, *22*(3), 404. https://doi.org/10.1093/NAR/22.3.404

Lefévre, C., & Lkeda, J.-E. (1994). A fast word search algorithm for the representation of sequence similarity in genomic DNA. *Nucleic Acids Research*, *22*(3), 404–411.

Li, H. (2016). Minimap and miniasm: fast mapping and de novo assembly for noisy long sequences. *Bioinformatics*, *32*(14), 2103–2110. https://doi.org/10.1093/BIOINFORMATICS/BTW152

Li, H. (2018). Minimap2: pairwise alignment for nucleotide sequences. *Bioinformatics*, *34*(18), 3094–3100. https://doi.org/10.1093/BIOINFORMATICS/BTY191

Limasset, A., Rizk, G., Chikhi, R., & Peterlongo, P. (2017). Fast and scalable minimal perfect hashing for massive key sets. *Leibniz International Proceedings in Informatics, LIPIcs*, *75*. https://doi.org/10.4230/LIPIcs.SEA.2017.25

López-Solano, A., Doadrio, I., Nester, T. L., & Perea, S. (2025). De novo genome hybrid assembly and annotation of the endangered and euryhaline fish Aphanius iberus (Valenciennes, 1846) with identification of genes potentially involved in salinity adaptation. *BMC Genomics*, *26*(1), 1–17. https://doi.org/10.1186/S12864-025-11327-0

Madeira, F., Madhusoodanan, N., Lee, J., Eusebi, A., Niewielska, A., Tivey, A. R. N., Lopez, R., & Butcher, S. (2024). The EMBL-EBI Job Dispatcher sequence analysis tools framework in 2024. *Nucleic Acids Research*, *52*(W1), W521. https://doi.org/10.1093/NAR/GKAE241

Magee, A. M., Aspinall, S., Rice, D. W., Cusack, B. P., Sémon, M., Perry, A. S., Stefanović, S., Milbourne, D., Barth, S., Palmer, J. D., Gray, J. C., Kavanagh, T. A., & Wolfe, K. H. (2010). Localized hypermutation and associated gene losses in legume chloroplast genomes. *Genome Research*, *20*(12), 1700–1710. https://doi.org/10.1101/GR.111955.110

Maizel, J. V., & Lenk, R. P. (1981). Enhanced graphic matrix analysis of nucleic acid and protein sequences. *Proceedings of the National Academy of Sciences of the United States of America*, *78*(12), 7665. https://doi.org/10.1073/PNAS.78.12.7665

Marino, S. M., & Gladyshev, V. N. (2010). Cysteine Function Governs Its Conservation and Degeneration and Restricts Its Utilization on Protein Surfaces. *Journal of Molecular Biology, 404*(5), 902–916. https://doi.org/10.1016/J.JMB.2010.09.027

Masutani, M., Sonenberg, N., Yokoyama, S., & Imataka, H. (2007). Reconstitution reveals the functional core of mammalian eIF3. *The EMBO Journal, 26*(14), 3373–3383. https://doi.org/10.1038/SJ.EMBOJ.7601765

Nacar, C. (2023). The Frequencies of Amino Acids in Secondary Structural Elements of Globular Proteins. *Clin Exp Health Sci, 13*, 261–266. https://doi.org/10.33808/clinexphealthsci.1239176

Nakaya, Y., Lilue, J., Stavrou, S., Moran, E. A., & Ross, S. R. (2017). AIM2-like receptors positively and negatively regulate the interferon response induced by cytosolic DNA. *MBio, 8*(4). https://doi.org/10.1128/MBIO.00944-17

Ndiaye, M., Prieto-Baños, S., Fitzgerald, L. M., Yazdizadeh Kharrazi, A., Oreshkov, S., Dessimoz, C., Sedlazeck, F. J., Glover, N., & Majidian, S. (2024). When less is more: sketching with minimizers in genomics. *Genome Biology 2024 25:1, 25*(1), 1–35. https://doi.org/10.1186/S13059-024-03414-4

Needleman, S. B., & Wunsch, C. D. (1970). A general method applicable to the search for similarities in the amino acid sequence of two proteins. *Journal of Molecular Biology, 48*(3), 443–453. https://doi.org/10.1016/0022-2836(70)90057-4

O'Cathail, C., Ahamed, A., Burgin, J., Cummins, C., Devaraj, R., Gueye, K., Gupta, D., Gupta, V., Haseeb, M., Ihsan, M., Ivanov, E., Jayathilaka, S., Kadhirvelu, V., Kumar, M., Lathi, A., Leinonen, R., McKinnon, J., Meszaros, L., Pauperio, J., … Cochrane, G. (2025). The European Nucleotide Archive in 2024. *Nucleic Acids Research, 53*(D1), D49–D55. https://doi.org/10.1093/NAR/GKAE975

Omelchenko, D., Bitja-Nyom, A. R., Matschiner, M., Malinsky, M., Indermaur, A., Salzburger, W., Bartoš, O., & Musilova, Z. (2024). Haemoglobin Gene Repertoire in Teleost and Cichlid Fishes Shaped by Gene Duplications and Genome Rearrangements. *Molecular Ecology, 33*(22), e17559. https://doi.org/10.1111/MEC.17559

Ordway, G. A., & Garry, D. J. (2004). Myoglobin: an essential hemoprotein in striated muscle. *The Journal of Experimental Biology, 207*(Pt 20), 3441–3446. https://doi.org/10.1242/JEB.01172

P. Bagowski, C., Bruins, W., & J.W. te Velthuis, A. (2010). The Nature of Protein Domain Evolution: Shaping the Interaction Network. *Current Genomics, 11*(5), 368. https://doi.org/10.2174/138920210791616725

Pearson, W. R. (2013). Selecting the Right Similarity-Scoring Matrix. *Current Protocols in Bioinformatics*, *43*, 3.5.1. https://doi.org/10.1002/0471250953.BI0305S43

Pop, M., Kosack, D. S., & Salzberg, S. L. (2004). Hierarchical Scaffolding With Bambus. *Genome Research*, *14*(1), 149. https://doi.org/10.1101/GR.1536204

Reisner, A. H., & Bucholtz, C. A. (1988). The use of various properties of amino acids in color and monochrome dot-matrix analyses for protein homologies. *Computer Applications in the Biosciences: CABIOS*, *4*(3), 395–402. https://doi.org/10.1093/BIOINFORMATICS/4.3.395

Roberts, M., Hayes, W., Hunt, B. R., Mount, S. M., & Yorke, J. A. (2004). Reducing storage requirements for biological sequence comparison. *Bioinformatics*, *20*(18), 3363–3369. https://doi.org/10.1093/BIOINFORMATICS/BTH408

Roeters, S. J., Golbek, T. W., Bregnhøj, M., Drace, T., Alamdari, S., Roseboom, W., Kramer, G., Šantl-Temkiv, T., Finster, K., Pfaendtner, J., Woutersen, S., Boesen, T., & Weidner, T. (2021). Ice-nucleating proteins are activated by low temperatures to control the structure of interfacial water. *Nature Communications*, *12*(1), 1–9. https://doi.org/10.1038/s41467-021-21349-3

Schmidt, E. E., & Davies, C. J. (2007). The origins of polypeptide domains. *BioEssays: News and Reviews in Molecular, Cellular and Developmental Biology*, *29*(3), 262. https://doi.org/10.1002/BIES.20546

Shoaei, F., Talebi-Ghane, E., Ranjbar, A., & Mehri, F. (2024). Evaluation of antibiotic residues in honey: a systematic review and meta-analysis. *International Journal of Environmental Health Research*, *34*(2), 1064–1075. https://doi.org/10.1080/09603123.2023.2197285

Shrestha, A. M. S., Frith, M. C., & Horton, P. (2014). A bioinformatician's guide to the forefront of suffix array construction algorithms. *Briefings in Bioinformatics*, *15*(2), 138–154. https://doi.org/10.1093/BIB/BBT081

Sigrist, C. J. A., De Castro, E., Cerutti, L., Cuche, B. A., Hulo, N., Bridge, A., Bougueleret, L., & Xenarios, I. (2013). New and continuing developments at PROSITE. *Nucleic Acids Research*, *41*(Database issue). https://doi.org/10.1093/NAR/GKS1067

Singh, V. K., Singh, P., Karmakar, M., Leta, J., & Mayr, P. (2020). The Journal Coverage of Web of Science, Scopus and Dimensions: A Comparative Analysis. *Scientometrics*, *126*(6), 5113–5142. https://doi.org/10.1007/s11192-021-03948-5

Smith, T. F., & Waterman, M. S. (1981). Identification of common molecular subsequences. *Journal of Molecular Biology, 147*(1), 195–197. https://doi.org/10.1016/0022-2836(81)90087-5

Sonnhammer, E. L. L., & Durbin, R. (1995). A dot-matrix program with dynamic threshold control suited for genomic DNA and protein sequence analysis. *Gene, 167*(1–2), GC1. https://doi.org/10.1016/0378-1119(95)00714-8

Staden, R. (1982). An interactive graphics program for comparing and aligning nucleic acid and amino acid sequences. *Nucleic Acids Research, 10*(9), 2951. https://doi.org/10.1093/NAR/10.9.2951

Stoltzfus, A., & Norris, R. W. (2015). On the Causes of Evolutionary Transition:Transversion Bias. *Molecular Biology and Evolution, 33*(3), 595. https://doi.org/10.1093/MOLBEV/MSV274

The UniProt Consortium. (2025). UniProt: the Universal Protein Knowledgebase in 2025. *Nucleic Acids Research, 53*(D1), D609–D617. https://doi.org/10.1093/NAR/GKAE1010

Wan, Q. H., Zeng, C. J., Ni, X. W., Pan, H. J., & Fang, S. G. (2009). Giant panda genomic data provide insight into the birth-and-death process of mammalian major histocompatibility complex class II genes. *PLoS ONE, 4*(1). https://doi.org/10.1371/JOURNAL.PONE.0004147

Wang, Y., Zhang, H., Zhong, H., & Xue, Z. (2021). Protein domain identification methods and online resources. *Computational and Structural Biotechnology Journal, 19*, 1145. https://doi.org/10.1016/J.CSBJ.2021.01.041

Wolber, P., & Warren, G. (1989). Bacterial ice-nucleation proteins. *Trends in Biochemical Sciences, 14*(5), 179–182. https://doi.org/10.1016/0968-0004(89)90270-3

Yates, A. D., Allen, J., Amode, R. M., Azov, A. G., Barba, M., Becerra, A., Bhai, J., Campbell, L. I., Carbajo Martinez, M., Chakiachvili, M., Chougule, K., Christensen, M., Contreras-Moreira, B., Cuzick, A., Da Rin Fioretto, L., Davis, P., De Silva, N. H., Diamantakis, S., Dyer, S., ... Flicek, P. (2022). Ensembl Genomes 2022: an expanding genome resource for non-vertebrates. *Nucleic Acids Research, 50*(D1), D996–D1003. https://doi.org/10.1093/NAR/GKAB1007

Zheng, H., Marçais, G., & Kingsford, C. (2023). Creating and Using Minimizer Sketches in Computational Genomics. *Journal of Computational Biology, 30*(12), 1251–1276. https://doi.org/10.1089/CMB.2023.0094

www.ingramcontent.com/pod-product-compliance
Lightning Source LLC
Chambersburg PA
CBHW061223240926
51846CB00006B/1329
9789199033013